animals&men

THE JOURNAL OF THE CENTRE FOR FORTEAN ZOOLOGY

I0202493

Issue 48

Dedicated with love to Biggles (2008-2010)

Typeset by Jonathan Downes,
Cover and Layout by SPiderKaT for CFZ Communications
Using Microsoft Word 2000, Microsoft , Publisher 2000, Adobe Photoshop CS.

First published in Great Britain by CFZ Press

CFZ Press
Myrtle Cottage
Woolsery
Bideford
North Devon
EX39 5QR

ISBN: 978-1-905723-65-2

EDITORIAL

Dear Friends,

This is a mildly embarrassing editorial for me to write. The vast majority of this issue was finished back in June, and should have appeared well before the Weird Weekend. The fact that it didn't is pretty well all my fault. Although nobody, especially me, is going to argue against the fact that this new format for *Animals & Men* is far more impressive than any of its predecessors, is also far more expensive, and it costs over £500 to publish and distribute each issue. Unfortunately, mathematics not being my strong point, I managed to totally screw up my calculations when costing out the new subscriptions, and our cash flow - as it is at the moment - is pretty much inadequate when it comes to ensuring that *Animals & Men* actually appears four times a year, as it is supposed to.

However, in mid-September we did something we probably should have done many years ago, and became a Company Limited by Guarantee. The Centre for Fortean Zoology, CFZ Press, and *The Amateur Naturalist* plus all the other things we do within the realms of both Cryptozoology and Natural History are now administered by a Company Limited by Guarantee called "CFZ Natural History Ltd". A Company Limited by Guarantee is basically

a limited company set up not to make a profit, and is the next step to being a charity. There are two Directors: Graham and myself, and - by law - neither director can benefit financially from the activities of the company. This has very few disadvantages, and quite a few advantages, and not the least being that we are now legally allowed to fundraise using methods that we could not before. Hopefully, this means that *Animals & Men* will once again have a publishing schedule that is something that one can rely upon. It is embarrassing to note that the last time we actually managed to produce four issues in a year was seven years ago.

An awful lot has happened in the last eight or nine months since we published *Animals & Men* 47. Some of it is covered in this issue, and other things such as the Weird

"THE GREAT DAYS OF ZOOLOGY ARE NOT DONE"

Weekend, our appearances at the *Fortean Times* Unconvention, and the current (as we go to press) expedition to India will be held forward until issue 49. I will do my best to guarantee that the next issue will not take another eight months to appear.

However, here is a brief roundup of news from the last eight months. All these stories will be revisited in more depth in future issues of *Animals & Men.*

GOODBYE BIGGLES

One of the saddest, if not the saddest event of the last year was the untimely, and completely unexpected death of Biggles, our two-year-old border collie. It turned out that he was almost certainly born with a congenital defect of the liver which none of us had suspected. His death was quick, and the news was devastating both to us, and his fans across the world. The CFZ is truly a global family, and Corinna and I took great comfort from the dozens of letters of support that we received. We are at present fostering a doggie, a five-year-old boxer x bulldog cross called Prudence who is in poll position to become the CFZ Doggie MkIV. It is a bit like *Doctor Who*. When one dog leaves us, it is never more than a few weeks before his or her successor ambles into our lives and takes possession of our hearts.

BIG CAT DNA

On the 17th August I was happy to announce that Danish zoologist Lars Thomas had examined hair samples found in Huddisford Woods near Woolsery, and pro-

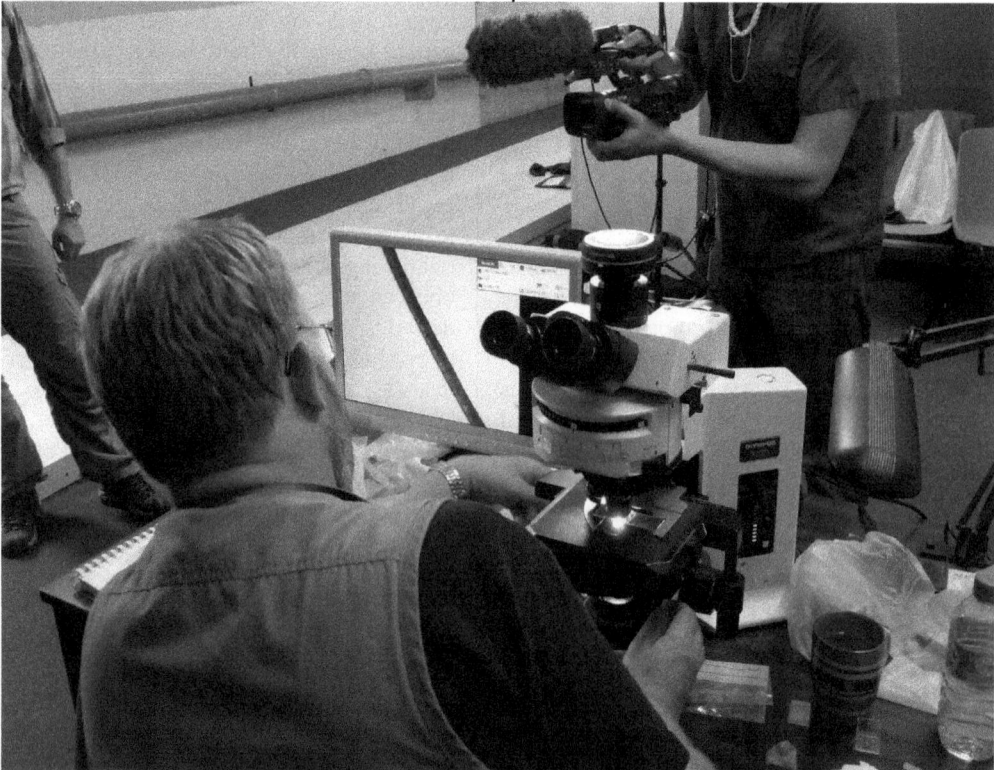

nounced them to be leopard.

I offered the hairs (which were found by Lars, Jon McGowan and a team known as the Four-Teans) to any research group or academic institution who wanted to try and verify Lars's findings. The first person to contact me was Dr Ross Barnett from Durham University who has done DNA analysis on them, and has confirmed that they are Pantherine, probably leopard.

He is carrying out further tests to establish the species and subspecies for certain and a full announcement will be made then.

BRITISH LYNX SPECIMEN

Another remarkable recent discovery was made in the archives of the Bristol City Museum and Art Gallery by Max Blake, one of the best known of the younger brigade of CFZ

members, and as I have said on a number of occasions, someone who together with Dave Braund-Phillips will be managing the circus once I've finally retired or died.

Although there have been reports of unknown cat-like animals in the UK on and off for centuries, there has been a paucity of hard evidence. Until now the only reports of specimens actually being secured are from the last 40 years or so. Max has discovered an animal, which appears to be a Canadian lynx, which was shot over a century ago.

He is currently working on a technical paper with Dr Darren Naish, which will contain all the details. He has asked us not to reveal any more details until it is published, and we of course agreed. When we are able to do so we shall have more photographs, and all the technical details that anyone could possibly want. Well done, mate.

POTENTIAL ORANG PENDEK DNA

A team of Danish scientists who have been analysing hair samples brought back from Indonesia by our expedition to Sumatra last autumn have found some potentially world-shattering results. The expedition was looking for the fabled orang pendek, an upright walking ape from Sumatra which is only known from eyewitness reports.

Expedition leader Adam Davies has been to Sumatra five times since 1999, to look for the orang pendek. Over the years, there has been a gradual refinement in his search technique. He is certain that it exists, and when he first went to Sumatra he was struck as to how authentic the first-hand accounts seemed to be. On a previous expedition in 2001, prints and hair were found, and subsequently examined by world famous hair analysis expert Professor Hans Brunner and by Dr David Chivers of Cambridge University. They independently concluded that they were from an unknown primate closely related to the two species of orang-utan.

At the Weird Weekend Lars Thomas announced the results so far. The preliminary DNA analysis of the hairs appears to resemble that of an orang-utan. He says:

"... the significance is quite enormous no matter what the result is basically, because if it turns out to be orang-utan this proves that there is orang-utan in a part of Sumatra several hundred kilometres from the nearest population of orang-utan. If it turns out to be a primate that looks like an orang-utan but isn't, it's an even greater discovery because that proves that there is another great ape living in Indonesia".

A morphological analysis of the hair samples also corroborated Professor Brunner's findings.

Richard Freeman, the zoological director of the CFZ has been to Sumatra on three occasions, the hairs in question being found on the last expedition in September 2009. On this particular trip were Adam Davies (leader), Richard Freeman, Chris Clark, Dave Archer plus their guides Sahar, John, Dally and Doni. It was the brother – John Didmus - of their main guide Sahar, who found the hairs on a small sapling about 3 feet off the ground. Richard said that:

"if the hair turns out to be from a new species, it would be the first confirmed upright walking ape which then throws an interesting light upon other reported bipedals like the yeti, etc. It may also help tell us how bipedalism in humans first developed. Also, the fact that such a large animal was found on an island roughly the same size as Britain could be significant as it may also mean that there could be other large animals still to be found across the world."

BIGFOOT FORUMS

In August we were approached by our old friend Paul Vella asking if we would like to rescue the Bigfoot Forums. They were, apparently, in danger of imminent closure.

After a brief discussion we agreed, and since then have been paying for and administering what are apparently the largest forums on the subject on the net. After the new forums had been online for some weeks, I wrote this open letter to the members:

Dear friends,

Now the revived Bigfoot Forums have been up for some weeks, and the dust has settled, and – to my great delight – a large number of you have relocated from the old forums to the new ones I think that it is important for me to introduce myself, and more importantly introduce the Centre for Fortean Zoology (CFZ).

For those who have not heard of us, the CFZ is an international organisation that I founded in the spring of 1992 to act as a clearing house for information on research into mystery animals across the board. Since 1994 we have been publishing books and magazines, since 1997 we have been carrying out a programme of expeditions across the globe, and since 2006 we have been making our own films and promulgating them for free through our YouTube channel CFZtv.

Until now we have not been particularly involved in Bigfoot research. It is not that we are not interested in man-beasts, quite the contrary, but until now most of our research has been in Asia. Since 2003 we have carried out three expeditions to the Indonesian island of Sumatra in search of the semi-legendary upright walking ape orang pendek. We brought back hair samples from the most recent expedition and preliminary testing implies that they are from a creature related to the orangutans which is pretty cool, as there aren't any orangutans in that region of the island and haven't been in historic times.

When Paul Vella contacted me and suggested that we might like to take over hosting Bigfoot Forums I agreed because I think it is important for the good of the whole cryptozoological community that places like Bigfoot Forums continue to exist. This is not just because they provide forums for discussion and the sharing of ideas, but they are invaluable in fostering a sense of community and they also provide an invaluable archive for researchers across the world.

What is the CFZ going to do with Bigfoot Forums? Well nothing really; it is doing pretty well without us, and we are simply pleased that we are in the position to pay the hosting costs. However, we are also in a position to help. We can provide DNA analysis and hair analysis through our colleagues in Copenhagen, and would be happy to do so if an opportunity arose. We do not want to tread on anyone's toes, merely to help in the quest to solve one of the most enduring mysteries on the planet.

Members of the forum are welcome to join in discussions on our daily interactive blog, and – indeed – we would welcome contributions for this or any of our hard copy publications.

If any of you want to get involved with CFZ activities, please feel free, and likewise if there is anything that we can do to help your researchers our resources are open to you.

Here's to a long and happy relationship.

Jon Downes

INDIA EXPEDITION

As I type this, the India expedition is deep in the heart of the Garo Hills in northern India, on the track of the animal most usually known in the west as `The Indian Yeti`.

I think that I cannot do better than reprint the press release that was sent out to the British Press (and universally ignored):

On 31st of October the CFZ 2010 expedition leaves England. They will be exploring the Garo Hills in Northern India in search of the mande-burung or Indian yeti. The five-man team consists of team leader Adam Davies, Dr Chris Clark, Dave Archer, field naturalist Jonathan McGowan, and cryptozoologist Richard Freeman.

The creatures are described as being up to ten feet tall, with predominantly black hair. Most importantly, they are said to walk upright, like a man. Walking apes have been reported in the area for many years. These descriptions sound almost identical to those reported in neighbouring Bhutan and Tibet. Witnesses report that the mande-burung, which translates as forest man, is most often seen in the area in November.

The Garo Hills are a heavily forested and poorly explored area in Meghalaya state in the cool northern highlands of India. The area is internationally renowned for its wildlife, which includes tigers, bears, elephants and Indian rhino and clouded leopards.

The Indian team will be led by Dipu Marek, a local expert who has been on the trail of the Indian yeti for a number of years and has found both its nests and 19inch long `footprints` on previous occasions. The expedition team has also arranged to interview eyewitnesses who have seen the Mande-Burung.

Camera traps will be set up in sighting areas in the hope of catching one of the creatures on film.

The Mande-Burung may be a surviving form of a giant ape known from its fossilised teeth and jaw bones, called Gigantopithecus blacki, *which lived in the Pleistocene epoch around three hundred thousand years ago. This creature is of course extinct. However, much contemporary fauna such as the giant panda, the Asian tapir and the Asian elephant that lived alongside the monster ape, still survive today. It is thought by many that Gigantopithecus also survives in the impenetrable jungles and mountains of Asia. Its closest known relatives are the Orangutans of Sumatra and Borneo.*

So, I think that you will agree - so far 2010 has been a momentous year for the Centre for Fortean Zoology. I sincerely hope that next year is one in which we shall be able to keep up the impetus, but also one during which we manage to get our administration and cash flow under control.

Thank you to everyone who has supported us during the past months. They have not been easy ones here at the CFZ. They have been ones of great change as well as successes, and the stresses and strains do sometimes show.

The bloggo is going from strength to strength, and I would like to thank the regular contributors as well as the recently engaged Lizzy Clancy, and Gavin Lloyd Wilson who work jolly hard each day for no reward and little recognition. Thanks guys.

But most of all I would like to thank my darling wife Corinna for being my helpmeet and co-conspirator now and always. Without you, honey, I would not be able to do any of this. Thank you.

Until next time,
Onwards and Upwards

Jon Downes,
Director, Centre for Fortean Zoology

THE FACULTY OF THE CENTRE FOR FORTEAN ZOOLOGY

"In her abnormalities, nature reveals her secrets." (Goethe)

PERMANENT DIRECTORATE

Hon. Life President:
Colonel John Blashford-Snell

******

Director:
Jonathan Downes
Deputy Director:
Graham Inglis
Administrative Director:
Corinna Downes
Zoological Director:
Richard Freeman
Deputy Zoological Director:
Max Blake

******

Membership Secretary:
Corinna Downes
North American Membership Secretary:
Naomi West
Ecologist: Oll Lewis
Technical: David Braund-Phillips

Directorate Assistant: Liz Clancy
Local Services: Matt Osborne
Trainee: Ross Braund-Phillips

******

Big Cat Study Group:
Neil Arnold
Aquatic Monster Study Group:
Oll Lewis
BHM Study Group:
Paul Vella

*****

North American Office:
Nicholas Redfern
Australian Office:
Rebecca Lang and Mike Williams
New Zealand Office:
Tony Lucas

Tyneside:
Mike Hallowell
West Midlands:
Dr Karl Shuker
Raheel Mughal
Wiltshire:
Matthew Williams
Yorkshire:
Steve Jones
Mark Martin

Wales

Gavin Lloyd Wilson
Elliot Saunders
Oll Lewis
Gwilym Ganes

Northern Ireland

Gary Cunningham
Ronan Coghlan

USA

California:
Greg Bishop
Dianne Hamann
Illinois:
Jessica Dardeen
Derek Grebner
Indiana:
Elizabeth Clem
Michigan:
Raven Meindel
Missouri:
Kenn Thomas
Lanette Baker
New York State:
Peter Robbins
Brian Gaugler
New Jersey:
Brian Gaugler

North Carolina:
Shane Lea
Micah Hanks
Ohio:
Chris Kraska
Brian Parsons
Oklahoma:
Melissa Miller
Oregon:
Regan Lee
Texas:
Chester Moore
Naomi West
Richie West
Ken Gerhard
Nick Redfern
Wisconsin:
Felinda Bullock

International

Australia:
Tim the Yowie Man
Mike Williams
Paul Cropper
Rebecca Lang
Tony Healey
Denmark:
Lars Thomas
Eire:
Tony 'Doc' Shiels
Mark Lingard
France:
Francois de Sarre
Germany:
Wolfgang Schmidt
New Zealand:
Tony Lucas
Roo Besley
Peter Hassall
Switzerland:
Georges Massey
United Arab Emirates:
Heather Mikhail

CONTENTS

The Centre for Fortean Zoology is a non-profit making organisation administered by CFZ Natural History Ltd, (Company #7381545) a Company Limited by Guarantee.

From humble beginnings we have become the largest, and we like to think the best, cryptozoological organisation in the English-speaking world. We have published over 65 books, 48 editions of our journal *Animals & Men*, as of November 2010, 38 episodes of our monthly webTV show *On the Track*, 3 feature length films and dozens of shorter features, 11 annual conferences and over a dozen foreign expeditions.

With a track record like this, what can we hope to achieve by attaining charitable status? Surely it is not just about the money. Well of course, it is not *just* about the money, but money is a necessary evil. Whilst we are – we hope – justifiably proud of our achievements to date, if we manage to get extra cash in as a result of our attaining charitable status there are a number of projects that we want to fund. For example, we would like to provide laptops and software for various researchers who would otherwise not be able to afford them. We want to set up static educational displays featuring posters, models and live exhibits in schools, hospitals, hospices, pubs, and restaurants across the southwest. We want to expand our programme of educational activities, and we would like to widen the scope of our Outreach projects.

So keep your fingers crossed.

EDITED AND COMPILED BY CORINNA DOWNES

Welcome to the *Animals & Men* #48 Newsfile. There are quite a few of what my mother would call 'creepy crawlies' in this issue, but she would be relieved to know that there are no snakes (or 'slithery things'). As with the last issue, it was a difficult task compiling the list and squeezing it down to fit in the allotted space – some creatures have so much interesting information attached to their discovery, or indeed re-discovery - that it is always a shame to have to cut it out.

It still amazes me that in the 21st Century new species are still being found, and that others that were previously thought to be extinct re-emerge into our world, having hung on to existence, unbeknown to us, with 'tooth and claw'. It makes you wonder exactly how many more creatures are out there hiding in the undergrowth or the thickly canopied forests of this earth. I certainly await the next few months with bated breath to see what turns up in the news.

But back to this issue, I have to admit that my favourite headline has to be **'Extinct' Aussie frog didn't croak after all**.

Yes, a dreadful pun, but one that just had to be used given the subject matter.

New gecko species identified in West African rainforests

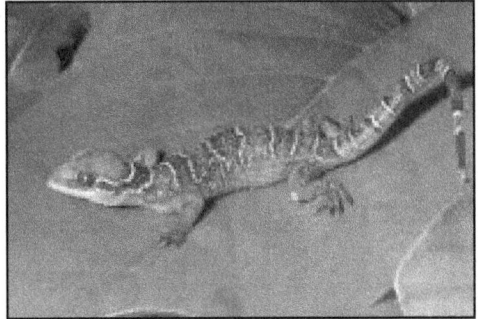

The West African forest gecko, is a secretive although widely distributed species in forests from Ghana to the Congo. It is actually four distinct species that appear to have evolved over the past 100,000 years due to the fragmentation of a belt of tropical rain forest, according to a report in a recent issue of the journal *Proceedings of the Royal Society B*.

"We tended to find this gecko, *Hemidactylus fasciatus*, throughout our travels in West Africa," said Leaché, a herpetologist with UC Berkeley's Museum of Vertebrate Zoology. "Despite the fact that it is recognized as one species, using new methods we have established a high probability that it is composed of at least four species."

"These rain forests are classified as one of the biodiversity hotspots on the planet, yet they are one of the most endangered areas on the earth," Leaché said. "Human deforestation is accentuating the process of habitat destruction."

Leaché and Fujita - who accompanied on several of the expeditions – decided to find out whether genetic diversity amongst the geckos could tell them something about the history of the rain forest belt. They found sufficient genetic differences among the 50 geckos collected from 10 different forest patches to identify four distinct species. The different species were found in different forest patches, suggesting that the species divergence was driven by the isolation of gecko populations from one another after gaps developed in the rain forest.

Not all of the species were separated by forest gaps, however. The wide Sanaga River in Cameroon is the dividing line between two species, which the researchers named *Hemidactylus coalescens* and *Hemidactylus eniangii*, the latter in honour of Nigerian conservation biologist and herpetologist Dr. Edem A. Eniang.

They retained the name *Hemidactylus fasciatus* for the westernmost species, which ranges from Sierra Leone to the wide Dahomey Gap, but identified an isolated species, *Hemidactylus kyaboboensis*, in the Togo Hills, which they named after Kyabobo National Park in the Volta Region of Ghana.

As a result, the researchers were able to state with high probability – essentially 100 percent that the specimens break down into four species.

SOURCE: 1ˢᵗ June 2010
University of California - Berkeley
Photo Credit: Charles Linkem

New giant monitor lizard discovered

According to a new study, a "spectacular" new species of giant, secretive, colourful and fruit-eating monitor lizard has been found in a Philippine forest. It is named the Northern Sierra Madre Forest monitor lizard *(Varanus bitatawa)*, and is 6 feet long, around 22 pounds

and brightly coloured yellow and black. It is in the same family as the Komodo dragon, the world's largest lizard.

"Rumours of its existence and some clues have floated around among biologists for the past 10 years," co-author Rafe Brown told *Discovery News*. He and his colleagues collected a large adult specimen from a forest at Northeast Luzon Island in the northern Philippines. They studied its anatomy and sequenced its DNA, both of which indicated that the lizard represents a new species. It is described in the latest issue of *Royal Society Biology Letters*.

"We think that it had not been discovered (before) primarily because of its secretiveness and because few comprehensive studies of amphibians and reptiles have been conducted in the inaccessible forests of NE Luzon Island," Brown said.

The huge lizard spends much of its time high in the trees overlooking the forest floor. Perhaps because of its size and apparent tree-specific body camouflage, it may be wary and cautious about exposing itself to terrestrial predators.

"We do not think it has a venomous bite," Brown said, thinking of the Komodo's venom. "It is not a carnivore, so it would gain no benefit from being able to deliver venom through its bite."

The researchers believe the animal is a "keystone species," which means it helps trees by eating their fruit. The seeds are prepared for germination after they pass through the lizard's digestive tract and are dispersed via bodily waste.

Eric Pianka, one of the world's foremost

experts on Varanus lizards, told *Discovery News* that this "new monitor lizard is indeed exciting. Who would have ever guessed that a 6-foot-long lizard could go undescribed until 2010?"

Although the lizard was undocumented until now, local Agta and Ilongot tribespeople have known about the animal. They rely on its meat as a major source of protein. Brown, however, thinks the greatest threats to the lizard's population are "deforestation, logging, mining and a lack of knowledge about biodiversity."

He and his colleagues have already collected specimens in the region representing at least another 10 species -- mostly lizards and frogs -- unknown to science.

"The Sierra Madre of Luzon is a treasure trove of undescribed vertebrate biodiversity," Brown said. "We suspect that many, perhaps dozens of new species of small vertebrates -- reptiles, amphibians, and possibly birds and mammals -- may await discovery in the forests of the northern Philippines."

SOURCE: 6th April 2010
Jennifer Viegas, *Discovery News*
Photo credit: Joseph Brown

MONITORING THE SITUATION

Varanus bitatawa is not the only new monitor to have been discovered this year. Two new monitors and one new sub-species – all from the Phillipines - were discovered after numerous preserved specimens were examined in various major European natural history museums and after long-term field studies. German scientist Andre Koch from the Zoological Research Museum, Alexander Koenig in Bonn, together with his supervisor Dr. Wolfgang Boehme and another colleague have described these species in *Zootaxa*.

Another new species closely related to the Kimodo dragon has been discovered in the Moluccan islands of east Indonesia. Known both as the torch monitor, and sago monitor, its scientific name is *Varanus obor* and is also a close relative of the recently reported fruit-eating monitor from the Phillipines. It can grow to nearly four feet long and lives on small animals and carrion and is found only on the small island of Sanana in the western Moluccan islands.

Sam Sweet (a professor in the department of Ecology, Evolution and Marine Biology at UC Santa Barbara), who along with Valter Weijola (a graduate student at Abo Akademi University in Turku, Finland) are the first to describe this lizard, "East of Wallace's Line -- the boundary between Asian and Australian domains -- there are no native carnivorous mammals, and monitor lizards fill that role. There are more species there, doing more different things ecologically than in Africa or South and Southeast Asia, where competition and predation by mammals tend to keep monitor lizards down. East of Wallace's Line in Indonesia, New Guinea, and Australia, monitor lizards are on the top of the heap. It emphasizes again how little we know about some tropical regions, to find an animal so strikingly colored and so large only last year."

New bird species found in rainforests of Borneo

Richard Webster – a Leeds University biologist - first glimpsed the bird from a canopy walkway 35m above ground. The spectacled flowerpecker - a small, wren-sized, grey bird - was feeding on some flowering mistletoe in a tree, and on one sighting it was heard singing.

The bird has white markings around its eyes, belly and breast. It has not yet been given a scientific name because so little is known about it.

Dr David Edwards, a tropical ecologist at the University of Leeds, identified the bird as a new species from photographs. "It's like a dream come true," he said. "I've spent all these years, decades, watching birds and all you want to do really is discover a new species to science. The team

caught sight of the birds several times in the days following its first appearance. "The discovery of a new bird species in the heart of Borneo underlines the incredible diversity of this remarkable area," said Adam Tomasek, leader of WWF's Heart of Borneo initiative.

The findings are published in the Oriental Bird Club's journal *Birding ASIA*.

SOURCE: 14th January 2010
Doreen Walton - Science reporter, *BBC News*
Photo credit: R.E. Webster

New bird discovered in Colombia – and released alive

Researchers have discovered a new species of antpitta in the Montane Cloud forests of the Colibri del Sol Bird Reserve in western Colombia. The antpitta is a thrush-like bird, and the new cinnamon and grey species was - according to a press release by the American Bird Conservancy (ABC) - "captured, banded, measured, photographed, sampled for DNA, and then released alive back into the wild".

The new bird has been named Fenwick's antpitta (*Grallaria fenwickorum*) after the President of ABC, George Fenwick, and his family, and is one of only a few incidences in which a new species has been described with-

out 'collecting' an individual (i.e. killing) to provide a model in a museum.

"I am deeply honoured by this naming. I know it reflects in equal parts on the contributions of both my family and the ABC organization, both of which have sought to further bird conservation efforts in Colombia," Dr. George Fenwick said in a press release. "I am especially pleased that this effort was achieved without the loss of the bird's life. Rare and special birds such as this should not have to be sacrificed to this process."

Colibri del Sol Bird Reserve, founded only five years ago, is managed by Fundación ProAves, a partner of ABC. The reserve is also home to the Critically Endangered (listed by the IUCN Red List) Dusky Starfrontlet that until 2004 hadn't been sighted for 50 years. Its rediscovery prompted the creation of the reserve.

Researchers have proposed that Fenwick's antpitta also be listed as Critically Endangered given that its population appears very small and much of the bird's original habitat has been cleared for pasturelands.

SOURCE: 26th May 2010
Jeremy Hance, *mongabay.com*
Photo credit: ©Fundacion ProAves

First record of greater kestrel in West Africa

June 2010 brought the discovery of a greater kestrel (*Falco rupicoloides*) in Termit (Niger) - the very first record for all of West Africa. The kestrel was spotted and photographed by project leader, Thomas Rabeil, and his team on February 16, 2010, during regular wildlife monitoring work. As it was unlike anything they had seen before, they sent photos of the bird to several world renowned bird experts, including Ron Demey and Nik Borrow, authors of the definitive *Field Guide to the Birds of West Africa*; Tim Wacher (*Birds of the Gambia and Senegal*); and Joost Brouwer and Ulf Liedén of the online Niger Bird Database. Greater kestrel was their unanimous response.

"We always knew Termit was a treasure house of Sahelo-Saharan biodiversity and an important overwintering and stopping off point for migrant birds from the palaearctic region but this amazing find goes to further underline its importance for biodiversity on an even larger scale".

A small to medium sized falcon, the greater kestrel - also known as the white-eyed kestrel – has a known range limited to the semi-arid lands of East and Southern Africa. The nearest known populations are in eastern Ethiopia. One can only speculate why and how it turned up several thousand kilometres from its known range. Birds of prey are known to migrate over vast distances and can be drawn into unfamiliar territory by storms or swarming locusts.

SOURCE: 2nd June 2010
John Newby, Sahara Conservation Fund.
Wildlifeextra.com

25 new beetles found in Turkish oaktrees - they prove hugely rich in species

It is known that old hollow oak trees provide important habitat for a large number of insects, especially beetles, and during a study of oak trees on grazing land in Turkey, 25 new species of beetle have been found.

"Most of them would disappear if the trees were to be cut down, and the risk is great", says project leader Nicklas Jansson, beetle ecologist at Linköping University (LiU) in Sweden.

There are 18 species of the oak family, *Quercus* in Turkey, and Nicklas Jansson and his co-workers from two Turkish universities have spent five years collecting beetles from oak trees in four large pastures in the south of the country near the border with Syria. These areas, 1,200-1,500 metres above sea level, are important for sheep and goat farming, but are now threatened by felling to make way for productive forest management.

Most of the newly discovered beetles belong to the Elateridae and Tenobrionidae families and have been identified by some 20 specialists across Europe.

The insects are collected by two types of traps, one which is mounted up in the trees to catch flying insects and one which is buried in the wood mould. The collection is done by students from the local universities.

"I hope that in finding new and unique species we will get the Turkish forestry authorities to open their eyes to their oak treasures and to begin conservation work in the most valuable areas," says Nicklas Jansson.

SOURCE: 2nd June 2010
WildlifeExtra.com

New species discovered in Indonesian mountain region

An international team of researchers was

camping in the Foja mountains of Indonesia - in the western side of the island of New Guinea - when herpetologist Paul Oliver spied a frog sitting on a bag of rice in the campsite. On closer look it turned out to be a previously unknown type of long-nosed frog. The scientists dubbed it Pinocchio. When the frog is calling, its nose points upward, but it deflates when the animal is less active.

"We were sitting around eating lunch," recalled Smithsonian ornithologist Chris Milensky. Oliver "looked down and there's this little frog on a rice sack, and he managed to grab the thing."

"Herpetologists (experts in snakes, lizards etc.) have good reflexes," Milensky observed. "He also caught a gecko; he managed to just jump and grab the thing" off a tree.

Overcoming torrential rain and floods, the researchers report finding the smallest kangaroo yet, a big woolly rat, a three-toned pigeon and a gargoyle-like, bent-toed gecko with yellow eyes.

So the environmental group Conservation International, with the support of the National Geographic Society and Smithsonian Institution, began investigating the area.

Kristofer M. Helgen, curator of mammals at

the Smithsonian's National Museum of Natural History, said one of the most amazing animals the researchers observed was the rare golden-mantled tree kangaroo. (left)

Most people think of kangaroos as creatures that live on the flatlands of Australia, he said, but this one has adapted to forest life.

"It can jump into a tree and scurry right up it," Helgen said. "But on the ground it hops around like any kangaroo."

While that kangaroo had been observed, rarely, before, Helgen also discovered what may be the smallest known member of the kangaroo family, a tiny wallaby that also has adapted to forest life. New Guinea and Australia were once connected and so have similar life forms, but they have adapted differently in each place, he explained.

A feature on this expedition appears in the June issue of *National Geographic* magazine.

"While animals and plants are being wiped out across the globe at a pace never seen in millions of years, the discovery of these absolutely incredible forms of life is much needed positive news," Bruce Beehler, a senior research scientist at CI and participant on the expedition, said in a statement.

"Places like these represent a healthy future for all of us and show that it is not too late to stop the current species extinction crisis," he said.

SOURCE: 17th May, 2010
Bloomington (IN) *Herald Times:* Randolph E. Schmid, *Associated Press*

Experts astounded by 'city of gonads' jellyfish

Tasmanian scientists have discovered a new species of jellyfish while surveying the waters outside the CSIRO in Hobart. The species was found in the River Derwent and is only a few millimetres wide and looks like a flying saucer with a cluster of gonads, or sex organs, on top. It has been named *Csiro medusa medeopolis*, meaning "jellyfish from CSIRO" and "city of gonads".

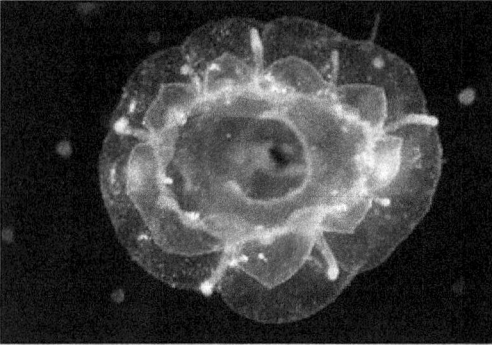

Launceston jellyfish expert Lisa-Ann Gershwin says it is an astounding discovery. "It's absolutely different from every other jellyfish that's ever been known," Dr Gershwin said. "So we not only put it into its own new species and its own new genus, but it's actually a brand new family."

Dr Gershwin says the find is also tremendously exciting. "Quite possibly and quite humbly the greatest discovery of my career, ever. I mean I'll be lucky if I ever get a discovery even half as incredible again. "You know any mum with a new bub is always excited, but when you have a whole family of new bubs I think it's triply exciting." Dr Gershwin says the jellyfish is harmless to humans.

SOURCE: 6th May 2010
ABC News
Photo credit: *ABC News*

New British moth found in Hembury Woods, Devon, UK is a world's first

Amateur naturalist Bob Heckford first spotted the tiny micro moth caterpillars in 2004 – they were an unusual bright green. In January this year the moth was officially recognised as a new species (*Ectoedemia heckfordi*). It has a wing span of just 6mm and body length of just 3mm and is not known to live outside the UK.

Mr Heckford is presenting the Natural History Museum with the original specimen, which is an important thing to do, because it marks the official acknowledgement by the scientific world of the specimen as the "type" for that species, against which any future finds will be compared and determined. "We hear so much about the losses to the natural world, and less about the gains; which makes this find, however small, so important," says Matthew Oates, an adviser on nature conservation at the National Trust.

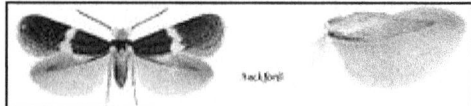

"Amateur naturalists have a wonderful window on the wildlife world and nature continues to amaze us and throw up surprises even in the UK."

Caterpillars of the new species are found mostly on oak saplings and low growth of oak in the shade. The mines they make are quite dark and the caterpillars are bright green which is quite unusual for micro moths. The adults lay their eggs on the underside of the leaf.

SOURCE: 28th April 2010
Matt Walker - Editor, *BBC Earth News*

Alcathoe's bat discovered in Yorkshire and Sussex

Researchers from Leeds and Sheffield universities have identified a species of bat that has never been seen before in the UK. The bats (*Myotis alcathoe*, or Alcathoe's bat) – about the size of the end of thumb - were found in woodland in Ryedale in the North Yorkshire Moors National Park and the South Downs in Sussex.

The bats were found during a Europe-wide study of bat population ecology and genetics, and it is believed they could be present in other parts of Britain.

The species was discovered in Greece in 2001 and is a native of continental Europe, but until now it was presumed that the English Channel had acted as a barrier preventing it from reaching the UK. Researchers believe the bat has not been spotted before because its appearance is similar to other species.

Professor John Altringham, from the University of Leeds, said: "Most of the bats were captured as they entered underground 'swarming' sites, where bats gather to mate

before going into hibernation." He added that this new discovery took the UK species from 16 to 17.

Brian Walker, Forestry Commission wildlife officer for the North York Moors, said: "We have some incredibly rich bat habitats in North Yorkshire. It was only a few years ago that work locally helped to confirm that the common pipistrelle was actually made up of two different species."

SOURCE: 20th April 2010
BBC News

Rare Worthen's sparrow nest sites found in Mexico

Three new breeding sites of one of the world's rarest birds, the Worthen's sparrow of Mexico, have been found. This discovery in the states of Nuevo Leon and Coahuila confirms the bird needs desert scrub to breed. Such information may help conservationists formulate a plan to save the species from extinction.

Just a few hundred Worthen's sparrows survive, and until the latest discovery little was known about where and how the bird reproduces. The sparrow (*Spizella wortheni*) was discovered originally in the US, where just one bird was caught on 16th June 1884, near Silver City, New Mexico.

21

None have been captured since in the country and it is now assumed extinct. It used to range over much of the Mexican Plateau, but Dr. Ricardo Canales-del Castillo of the Universidad Autonoma de Nuevo Leon in San Nicolas de los Garza, Mexico says that it is now found in just a handful of places in the Mexican northeast.

It is one of the rarest sparrows in North America , and although flocks are sometimes seen in the winter, there is little known about its summer breeding and living habitat- information which is crucial to safeguarding its future. The researchers, deciding on searching out places where the bird may be hiding, found this difficult because of the sparrow's behaviour - Worthen's sparrows do not migrate, but they move on from nesting sites as soon as the breeding season ends.

So the researchers had to catch the birds in the act of egg-laying and rearing. They found nests at San Rafael in Galeana, Nuevo Leon and at La Carbonera in the same state. Other nests were also found at San Jose del Alamito just over the border into the state of Coahuila, but still within the La Soledad valley.

Overall, the researchers recorded 51 individual sparrows, which is half the 100 to 120 individuals that survive in the wild, according to the latest data held by the International Union for the Conservation of Nature (IUCN). "Our most optimistic estimate is that 500 individuals remain," Dr Castillo told the BBC. Details of the new breeding locations are published in the *Journal of Field Ornithology*.

SOURCE: 22nd March 2010
Matt Walker - Editor, *BBC Earth News*
Photo credit: UANL/FCB

World's longest bug and 'Ninja' slug discovered in Borneo

Scientists have revealed the discovery of species that have recently come out of hiding in the rainforests of Borneo. These include the world's longest known stick insect, a slug that shoots "love darts," and a colour-changing frog.

The new WWF report details the 123 newly identified species that have been discovered since February 2007 when the three countries that make up Borneo agreed to conserve 85,000 square miles (220,000 square kilometers) of tropical rainforest, designated as the Heart of Borneo (HoB).

That's a rate of discovery of three species per month. Previously, scientists have estimated that there are about 2 million known species of life on Earth, and anywhere from 5 million to 100 million species that remain undiscovered.

"As the past three years of independent scientific discovery have proven, new forms of life are constantly being discovered in the Heart of Borneo," said Adam Tomasek, who leads the WWF project.

Here's an introduction to the new gang:

Longest insect - Measuring more than 1.6 feet (0.5 metres) in length, the world's longest stick insect, called *Phobaeticus chani*, was found near Gunung Kinabalu Park, Sabah. So far, only three specimens of the species have been found, all in the Heart of Borneo.

Fiery snake - Kopstein's bronzeback snake (*Dendrelaphis kopsteini*) is about 5 feet

(1.5 meters) long. Its neck is bright orange, which fuses into an iridescent and vivid blue, green and brown pattern that extends the entire length of its body.

Colour-changing frog - called *Rhacophorus penanorum*, this small frog species, whose males grow to just 1.4 inches (3.5 centimeters), was discovered in Gunung Mulu National Park, Sarawak, in the Heart of Borneo. Also called the Mulu flying frog, the amphibian has a small pointed snout and is unusual in that the species has bright green skin at night but changes colour to display a brown hue during the day. Its eyes follow suit to change colour as well. And while the minute animal may not fly with the birds, it uses its webbed feet and aerodynamic flaps of skin on the arms and legs to glide from tree to tree.

Ninja slug - This green and yellow slug *(Ibycus rachelae)* was discovered on leaves in a mountain forest at altitudes up to 6,233 feet (1,900 metres) in Sabah, Malaysia. The slug sports a tail that's three times the length of its head, which it wraps around its 1.6-inch-long (4 cm) body as if a pet cat. In fact, its discoverers initially planned to name the slug *Ibycus felis*, after its feline inspiration. Instead, they named it after the girlfriend of

one of its discoverers, Menno Schilthuizen of the Netherlands Centre for Biodiversity 'Naturalis.' "The distinction between slugs and snails is not so strict in that part of the tropics, because most of the slugs, including the new one we described, are semi-slugs meaning they still have a shell but the shell is so small that it can't retract its body into it," Schilthuizen told LiveScience.

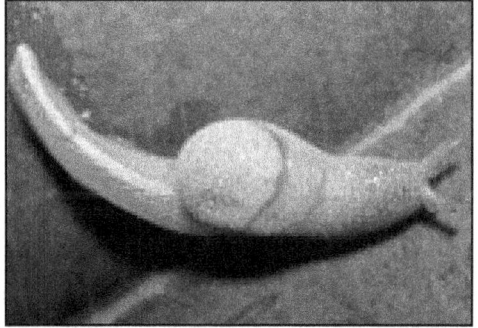

And though they've found several new slug and snail species, Schilthuizen said this rainforest environment isn't ideal for the animals. That's because the soil is highly acidic, which dissolves the animals' limestone shells. Overall, the Heart of Borneo is now called home by 10 primate species, more than 350 birds, 150 reptiles and amphibians and a staggering 10,000 plants that are found nowhere else in the world, according to the new report.

To keep these species and their lush home safe from demise, under the 2007 agreement, the three governments have committed to conserve and sustainably manage the area.

SOURCE: 21st April 2010
Jeanna Bryner *LiveScience* Managing Editor, livescience.com
Slug photo credit: © Menno Schilthuizen
Snake photo credit: © Gernot Vogel

'Extinct' Aussie frog didn't croak after all

When Luke Pearce, a fisheries conservation officer, stumbled across a frog in the Southern Tablelands of New South Wales he thought that it might be a rare yellow-spotted bell frog (a species thought to have been extinct for 30 years) and returned later with experts to try to identify it. David Hunter, a Threatened Species Officer, described how his legs were "trembling with excitement" as he approached the frog, and said he was "beside himself" when he checked the markings on its thigh which confirmed its identity.

The discovery was made last year, but secrecy has been maintained to protect the species, and experts plan to establish a colony in Sydney's Taronga Zoo, with the hope of reintroducing them into the wild at a later date once their numbers have grown. The species was thought to have been wiped out by a deadly fungus. But Dr Hunter says there are signs that the newly-discovered frogs have developed immunity against infections which should help them survive.

SOURCE: 4th March 2010
Ian Woods, Australia correspondent, Sky News

World's least known bird rediscovered

A species of bird, which has only been observed alive on three previous occasions since it was first discovered in 1867, has been rediscovered in a remote land corridor in north-eastern Afghanistan.

During the summer of 2008, American ornithologist Robert J Timmins was commissioned by the American aid organisation – USAID - to compile an inventory of bird species in the Badakshan province in north-eastern Afghanistan. He managed to record the call of a species of bird that was as yet unknown. This recording found its way to the Swedish ornithologist Lars Svensson, who was quick to note that the recorded birdsong did not resemble that of

any known species of bird. But from Timmins' description of the species, he soon began to suspect what kind of bird was on the recording.

Lars Svensson and Urban Olsson at the Department of Zoology, University of Gothenburg, had in fact shown in a previous study that about a dozen stuffed birds in museum collections all around the world had been incorrectly classified: they were not of the common species of reed warbler the curators had assumed, but rather a far rarer species known as the large-billed reed warbler (*Acrocephalus orinus*) - observed on just three documented occasions since 1867.

In their previous study Svensson, Olsson and co-workers had pinpointed North-Eastern Afghanistan as an area where the large-billed reed warbler probably bred in the 1930s.

When both the Swedish colleagues heard the recording of the mysterious birdsong they realised that they were on the trail of an ornithological sensation. News of the find was published in the journal *Birding Asia* and has aroused huge interest in ornithological circles. The large-billed reed warbler is not hunted, but is regarded as being under acute threat since its breeding sites are being deforested by the local population in their hunt for fuel.

"That's why it's vital that we protect both the species and its habitat now," says Urban Olsson.

Link to the article in *Birding Asia*: http://www.orientalbirdclub.org/publications/ba12pdfs/Timmins-LBWarbler.pdf

SOURCE: 19th January 2010
Krister Svahn, University of Gothenburg, Faculty of Science

Leopard cat found for 1st time in decades on Tsushima's lower island

Confirming their existence there for the first time in more than two decades, a rare leopard cat has been found for the first time on the lower island of Tsushima, in Nagasaki Prefecture, conservation officers have said.

Until recently, the highly protected Tsushima leopard cat, one of two species of wildcats found in Japan, was feared to have completely disappeared from Tsushima's lower island, though as many as 150 are thought to still survive on its upper island.

Officers of the Tsushima Wildlife Conservation Center said the 1,130- gram juvenile male - thought to have been born only last spring - was found in a weakened state on the property of a company in Izuhara town by an employee who notified the authorities.

Officials of the Tsushima Wildlife Conservation Center, established by the Environment Ministry to study the Tsushima leopard cat and assist in recovery of the critically endangered species, were summoned to the scene. Leopard cats leave their mother's home range 6 or 7 months after birth, at which time they must struggle to survive on their own. The Tsushima leopard cat, which is about the same size as a domestic cat but can be distinguished by a white spot on the back of each ear, is thought to have arrived on Tsushima from the Asian continent about 100,000 years ago.

Kamijima, Tsushima's larger and less populated upper island, is home to an estimated

80-110 of the small wildcats, down from an estimated 250-300 in the 1960s, conservation officers said. But on Shimojima, the lower island, the last confirmed wildcat sighting was in March 2007 when an automatic camera took a photograph of one, confirming their existence there for the first time since 1984 when one was found dead along a road, they said.

Wildlife officer Shinsuke Mizusaki said the leopard cat's numbers have been declining throughout Tsushima mainly due to habitat loss and road kill. Since 1991, 42 of the wildcats have been killed on Kamijima roads, including one earlier this month. To reverse the decline of the Tsushima leopard cat, which was designated by the Japanese government as a Natural Monument in 1971, it was declared a National Endangered Species in 1994 and a government-funded project was established to protect it.

The project involves field research, habitat restoration, captive breeding and public education about threats to the wildcats which also include diseases carried by domestic cats, illegal snare trapping and feral dogs. In recent years, the Japanese government has been studying the feasibility of reintroducing wildcats to Shimojima.

The Tsushima leopard cat, which goes by the scientific name *Prionailurus bengalensis euptilura*, is regarded as an isolated subspecies of the leopard cat, found across Eurasia. Japan's other wildcat species is the Iriomote cat, or *Prionailurus iriomotensis*, found on the island of Iriomote in southern Okinawa Prefecture.

Source: 29th December 2009
Nagasaki, *Associated Press*

DNA identifies new ancient human dubbed 'X-woman'

Scientists have identified a previously unknown type of ancient human through analysis of DNA from a finger bone unearthed in a Siberian cave. The tiny fragment of bone from a fifth finger was uncovered by archaeologists working at Denisova Cave in Siberia's Altai Mountains in 2008, and an international team has sequenced genetic material from the fossil showing that it is distinct from that of Neanderthals and modern humans. Details of the find, dubbed "X-woman", have been published in *Nature* journal.

The extinct "hominin" (human-like creature) lived in Central Asia between 48,000 and 30,000 years ago, and Professor Chris Stringer, human origins researcher at London's Natural History Museum, called the discovery "a very exciting development".

"This new DNA work provides an entirely new way of looking at the still poorly-understood evolution of humans in central and eastern Asia."

The discovery raises the intriguing possibility that three forms of human - *Homo sapiens*, Neanderthals and the species represented by X-woman - could have met each other and interacted in southern Siberia. An international team of researchers extracted mitochondrial DNA from the bone and compared the genetic code with those from modern humans and Neanderthals. The analysis carried out by Johannes Krause from the Max Planck Institute for Evolutionary Anthropology in Leipzig, Germany, and colleagues revealed the human from Denisova last shared a common ancestor with modern humans and Neanderthals about one million years ago.

"Whoever carried this mitochondrial genome out of Africa about a million years ago is some new creature that has not been on our radar screens so far," said co-author Professor Svante Paabo, also from the Max Planck Institute for Evolutionary Anthropology.

The divergence date of one million years is too young for the Denisova hominin to have been a descendent of *Homo erectus*, which moved out of Africa into Asia some two million years ago. And it is too old to be a descendent of *Homo heidelbergensis*, another ancient human thought to have originated around 650,000 years ago. However, for now, the researchers have steered away from describing the specimen as a new species.

It is particularly interesting for those of us who are looking for living hominins of an unknown species in the former Soviet central Asia.

SOURCE: 25th March 2010
Paul Rincon
Science reporter, *BBC News*

Pine martens make comeback in the UK after leading a secret life for decades

One of Britain's rarest and most elusive mammals - the pine marten – is back. And the reason is it never quite went away. A new report reveals pine martens are not confined to the fringes of the UK as was assumed, but that they have been living a secret life under our noses for decades.

Pine martens are related to weasels and otters and are agile, inquisitive, and about the size of a small cat. With their deep chestnut fur and yellow bib they are one of Britain's prettiest mammals. They are also supreme forest predators and hunt voles, rabbits, squirrels, birds and have a taste for honey, nuts, fruit and fungi.

The beautiful fur made their pelts valuable, and their killer instinct, particularly when it came to pheasants and partridges, made them enemies of gamekeepers. Pine martens were consequently persecuted to extinction in most parts of Britain but, even with full legal protection in 1988, only small enclaves hung on in remote parts of northern Scotland, Cumbria and north Wales. Or so it was thought.

The report from the Vincent Wildlife Trust, based on 12 years of research and sightings, reveals a surprise: pine martens are largely still present in the same parts of England and Wales from which they were recorded in the past. This includes areas such as Carmarthenshire, Montgomery, North York Moors and the Cheviots but also sightings in Cambridge, East Sussex and Northampton.

How did this 80cm-long nocturnal tree dweller pass unnoticed for so long? The report's author, pine marten expert Johnny Birks, told the *Guardian*: "Not everybody overlooked the pine martens. Some dedicated naturalists, who had been watching them for many years, kept hope burning that the species' presence would be recognised. The trouble is the authorities are used to cheap and easy survey methods which don't work on something as elusive and scarce as pine martens and the wrong perception has arisen that there were none left."

For the period 1996 to 2007, researchers analysed records, talked to people with convincing sightings, collected roadkill and used DNA testing on scat to slowly piece together the true picture.

The story also tells us how important the observations of amateur naturalists are, said Birks. "There are cultural changes towards instant results and fewer people dedicated to plugging away, building up their skills hoping to strike it lucky in pursuit of rare animals. These sorts of naturalists are as rare as the animals themselves." The "reappearance" of the pine marten is a further encouragement to create new woods and protect old ones, particularly old trees with breeding holes in them, said Birks, as well as championing greater connectivity between woodlands so pine martens can stay aloft and away from predatory foxes.

The new coalition government has pledged to launch a national tree planting campaign, with a target of a million trees. It is expected to begin in the autumn.

SOURCE: 4th June 2010 Paul Evans, guardian.co.uk
Photograph credit: Natural England/ PA

An alien Grey

A grey whale (*Eschrichtius robustu*) has been seen off Herzliya Marina, Israel - this species was thought to be extinct in the Atlantic Ocean. Although the appearance in the Mediterranean Sea is a surprise, it is thought that the whale may have travelled a vast distance from its natural habitat in the Pacific Ocean – thousands of miles away. Its appearance off the coast of Israel however has sparked the thought that grey whales may have returned to former haunts in the western hemisphere. There were once three major populations of this whale: in the western and eastern North Pacific and in the North Atlantic, the latter becoming extinct some time during the 17th or 18th Century.

"This discovery is truly amazing. Today, gray whales only inhabit the Pacific Ocean, so to find one in the North Atlantic, let alone the Mediterranean Sea, is bizarre in the extreme," says Nicola Hodgins of the Whale and Dolphin Conservation Society (WDCS), which has its headquarters in Wiltshire, UK.

Gray whales are well known for performing one of the world's longest migrations, making a yearly round trip of 15,000-20,000 km," says Ms Hodgins. "However, these new images show that this particular whale would have had to beat all previous distance records to end up where it has."

SOURCE: 10th May 2010
By Matt Walker - Editor, *Earth News*

XTRA NEWS FILE

[untitled]

When I started the Centre for Fortean Zoology in 1992 it was mostly to give validation to my own personal researches into the mystery animals of the West Country, but it was also because I felt that the nature of cryptozoology as a whole needed to be redefined. I have recently come across the term 'ethnoknown', first in a post on Cryptomundo, and secondly in Chad Arment's smashing new book on mystery carnivores of North America. I am surprised I have not come across the term before, and I suspect it is a word of recent origin. But it is a remarkably apposite term for cryptozoologists for cryptozoology does indeed concern animals that are less well-known; i.e. animals that are known to the inhabitants of an area (even if this knowledge is only in folkloric or zoomythological terms). This is, after all, the very essence of cryptozoology.

It has been claimed that although the CFZ have carried out over 20 expeditions they have come back with no evidence for the existence of any of cryptids for which they have searched. If you are examining these expeditions in cryptozoological terms, and we are of course cryptozoologists, this is just simply not true. Each expedition has come back with anecdotal evidence from the people 'on the ground' which bolster

what we know of these cryptids as ethnoknown, but still cryptic - as far as the scientific community on the whole - animals.

One of the things that I have always thought massively important as far as cryptozoology is concerned, and one in which has therefore become one of the watchwords of the Centre for Fortean Zoology as a whole, is that cryptozoology is not the study of monsters. Whereas the study of ethnoknown creatures can be used to extrapolate the existence of various lake monsters, man-beasts, and other fearsome denizens of far flung places, it can also be used far closer to home to extrapolate the existence of far less exciting but equally significant animals.

For example, in the first few issues of *Animals & Men* (now collected together as *Animals & Men Issues 1 – 5 In the Beginning* (CFZ 2001) and my own *Smaller Mystery Animals of the West Country* (CFZ 1996) I presented evidence for the existence of three ethnoknown mystery animals in the south-west of England.

They were:

- A British population of the green lizard (*Lacerta viridis*) in southern Dorset and southern Devon. I hypothesised that they could have become naturalised in the area after having been inadvertently introduced through the south coast seaports into which they had been imported in shipments of fruit and flowers from the Channel Islands in which this species has long been resident.
- Surviving populations of the pine marten (*Martes martes*). This charming little carnivore was, according to Langley and Yalden writing in 1977, extirpated from its entire English range by

the end of the 19th Century. In *The Smaller Mystery Carnivores of the West Country* I presented evidence that in Devon, Cornwall, Somerset, Dorset and Hampshire, and possibly Surrey, this was just not so and that (in some areas bolstered up by artificial and unofficial introduction programmes) this was just not true.

- Working on evidence from 15th, 16th and 17th Century parish records I concluded that Edward Alston's 1872 paper on the '*Specific Identity of the British Marten*' was quite simply wrong. In almost all of its range, *M.martes* co-exists with another – and closely related – species: *Martes foina*. The beech or stone marten, or marten cat as it is commonly known, was thought to live in various parts of the UK and Ireland until – with a stroke of the pen – Alston disenfranchised it. Such things happen all the time in zoology.

For example, in Hong Kong, the one cervid was thought to be the Chinese barking deer, Reeve's muntjak (*Muntiacus reevesi*) and was included in all the reference books as such together with photographs which are undoubtedly of this animal. Half way through the last decade there was a paradigm shift and suddenly the word from on high was that the only deer living in the former British colony was the Indian or red muntjak (*Muntiacus muntjak*). I believe that both species, and quite probably hybrids of the two in various degrees of introgressiveness exist there. But this is, of course, another story.

Over the 16 years which have transpired since I first published these theories, I have – to a certain extent – been vindicated on two of them.

- Firstly the green lizards. Sometime during the 1990s *Lacerta viridis* was split into several species and the ones found in western Europe are now called *Lacerta bilineata*. The great surprise in the world of herpetology but not to those of us who follow Heuvelmans' suggested methodology *vis a vis* ethnoknown animals, populations of *Lacerta bilineata* were discovered near Bournemouth – a sea port with a regular congress to and from the Channel Islands. The latest accepted thinking is that these animals are of relatively recent introduction, however whether they were introduced deliberately or by accident using a model similar to the one suggested by me back in the halcyon days of 1994 remain – for the moment – obscure. However, I think that I have been fully vindicated and strongly expect similar colonies to be found in the hills above Seaton and Lyme Regis in the next few years. Possibly I wasn't justified in the ungentlemanly headline "Told u so!" that I used both in *Animals & Men* and the late-lamented *Pet Reptile* magazine, but "I was so much older then, and younger than that now".

- The most recent vindication of my personal use of the study of ethnoknown creatures in the UK came in the *Guardian* on 4th June 2010 when a report by the quango Natural England was discussed. It turns out that far from pine martens being extinct, small pockets of survivors have hung on quite successfully in various parts of England and Wales, as well as the areas in Scotland and Ireland where it has always been known to be living. Unfortunately for those who would like to see me hailed as some sort of zoological

hero, the furthest south that the Natural England report said that pine martens had been seen was Northamptonshire, but I never particularly wanted to be a hero anyway. Again, I think that it almost certain that I should be proved 100% correct in the next few years. But when this happens, it will be a victory for cryptozoology, not a victory for the CFZ, and that is how it should be.

So what about my third prediction of 1994? Will beech martens be found to be UK residents? I still think, almost certainly, yes. The British list of vertebrates is far from being written in stone. In the last 15 years the pool frog (*Rana (Pelophylax) lessonae*), and at least two species of bat - Alcathoe's bat (*Myotis alcathoe*), and Nathusius' Pipistrelle (*Pipistrellus nathusii*) – have been proven to be British residents. The latter bat, by the way, was one that I had hypothesised was a British resident back in 1992.

I think that whilst the question of the existence of Bigfoot, the Loch Ness monster, and the yeti, and even the existence of big cats in Britain may still be questionable a decade from now, the three claims that I made in issue 1 of *Animals & Men* have certainly been vindicated and will be proof that cryptozoology, as laid down by Heuvelmans himself, is - indeed - a valid discipline because all three of my predictions were based on eyewitness testimony. All three creatures were ethnoknown as British residents.

This, I think, gives all of us who are interested in such things, great hope for the future. The astute amongst you will notice that there isn't a proper title to this piece. When he came to tea recently, Dr. Naish repeated his assertion that the title of my 2002 article in Tet Reptile magazine was probably a mistake if I was

trying to be taken seriously by workers in the scientific establishment. I probably should not invite Dr. Naish to tea so often because my lovely wife decided to take Dr. Naish's part in this disagreement.

Furthermore, she wouldn't let me entitle this present paper 'In Yer Face you big bunch of @*&^', and so together with my quote from 'my back pages' I must let you all know with my new title that I have been Byrd watching.

DISCLAIMER

Animals & Men is published by a team of unpaid volunteers, and so we have to admit that sometimes we are not able to do as much work as we would be able to if we had a team of paid office staff. One area in which we admit that we have been lax in the past is picture clearance. Whilst every effort has been made to contact copyright holders of the images used in the news sections of this journal, we will be the first to admit that occasionally something slips through the net.

Animals & Men is a not-for-profit publication, and in our opinion all images we use are covered by `Fair Use` legislation. However, if you are the copyright holder of an image that we have used without permission, please feel free to contact us at the editorial address, and we shall make every effort to resolve the situation amicably.

OBITUARY

I was working on Karl Shuker's new book over the weekend when I decided on a whim to see to whom this latest volume was dedicated. I was very shocked when I read:

"...And also to the memory of Jan Williams, who was not only a highly proficient and totally professional cryptozoological researcher but also a good friend and a lovely lady."

She was, indeed, a lovely lady, and she was my friend as well, although I am mildly embarrassed to admit that we drifted apart over ten years ago. However, if it had not been for her help and kindness in the early days, the CFZ as we know it would never have come into existance. As I write in my autobiography:

I have to admit, with hindsight of a over a decade, at that time I was becoming very frustrated with my cryptozoological research. I had been doing it as a hobby for nearly 25 years, and whilst it was all very well forming something called the Centre for Fortean Zoology, I had no idea how I was going to turn my vision into reality. So, when during the summer of 1993 one of my acquaintances - a lady called Jan Williams who lived in Congleton in Cheshire - announced that she was going to start up her own cryptozoological organisation, I was happy to put my own plans for the CFZ on hold and throw myself into working for S.C.A.N - The Society for Cryptozoology and the Anomalies of Nature.

The arrival of this new organisation could not have come at a better time for me, because although I was quite happy to continue my researches, pressure of other commitments was getting in the way of my plans for starting up the ultimate Cryptozoological research organisation of my own.

And half a chapter later...

Back in the world of cryptozoology, my new-found position as a member of the rank and file of SCAN was not going too well either. I have never found out why - and after this length of time it is none of my business - but all was not well [.....] After only four issues of their newsletter she announced that they were going to close. Then, I had a wonderful idea. I telephoned Jan and suggested that she joined me in making my vision of the Centre for Fortean Zoology into a reality. I told her of my background in Small Press Publishing

JAN WILLIAMS

and suggested that we start a magazine dedicated to cryptozoology. I even had a name for it - *Animals & Men* (the name of a song on the first album by *Adam and the Ants*). I suggested to her that if we were to take over the membership list of SCAN (after having let all the members have the chance of their money back if they wanted it), then we would not be in the awkward position of starting up a new publication without the benefit of having any readers for it. To my great joy, Jan agreed, and in April 1994 [...] the first, faltering steps towards a proper Centre for Fortean Zoology had just been made.

So, although I hadn't spoken to Jan, her husband Keith (who incidentally coded our first proper website) and their kids for something like 12 years (Michael, the little boy who did drawings for one of the earliest issues of A&M must be in his twenties now, and his younger sister was a toddler last time I saw her but must be in her late teens), the news that she died of cancer earlier in the year was an enormous shock. I just want to say to Jan: thank you, my dear. Without you my life would have been immeasurably different. I owe you a debt that can never be repaid.

WE ALSO SAY GOODBYE TO.....

This issue we also say goodbye to Nick Redfern's mum who died this summer after a long illness.

I only met her the once, about ten years ago, but she was a lovely, intelligent, and vivacious lady who made a deep impression on me. She will be sadly missed, and the hearts of the CFZ, collectively, and Corinna and I specifically go out to Nick and Dana Redfern and his father Frank.

(Sir) Laurence Gardener

On 14th August Helen Sanderson wrote to me:

Hi, Jonathan, This just posted by a friend, and apologies for being the bearer of bad news, since I know he was a friend of yours, though perhaps because of that you already know...Message:

'It is with great sadness that I write this announcement that my wonderful friend Laurence Gardner passed away yesterday after a long illness in the early morning hours. Laurence was a brilliant man and author of such best selling books as *Bloodline of the Holy Grail, Genesis of the Grail Kings, The Realm of the Ring Lords* and many more extraordinary works. Etc..'

Sorry.
Helen

I hadn't seen Laurence for some years, after he took exception to a review of one his books written by a friend of mine. He goes down in history as being the first person ever to threaten us with legal action!

However, although I thought that 90% of what he wrote was bosh and doubted the legitimacy of his title, which was granted by a Belgian geezer who claimed to be the Stuart pretender to the British throne, I did indeed quite like him!

It seems appropriate, for someone whose life was so full of strangeness and obfuscation that Wikipedia refused to post a notice of his death for some weeks because someone had cast doubt upon the event and there was no notice of it on his website.

AQUATIC MONSTERS LOG BOOK

BY OLL LEWIS

Nessie Dead Again

On a slow news-day this January the U.K. and international press started to publish a story with the shocking news that Nessie was dead! This would have been a lot more shocking, however, if they had not been banging this same drum several times for at least the past three years. The evidence for the passing of Nessie was that there had only been one or two credible sightings of the monster in the press in the last year. The fact that there are fewer reports published in the press has a lot less to do with the existence/continued existence of Nessie than it does with the times we live in.

There is a culture of ridiculing the strange in Britain today thanks in part to the views of certain scientists, who don't follow the scientific method getting held up on pedestals by the media.

The media will often give a scientist with an agenda a documentary where they sit on the banks of Loch Ness, or some other place where cryptids have been sighted, for half a day and when they see nothing they will proudly proclaim that they have solved the mystery by disproving the existence of anything strange because they didn't see it in a few hours. Sadly, as TV shows like this are the only exposure most people have to scientists they think that chumps like this speak for all of science as a whole and science says there is no such thing as cryptids.

Of course nothing could be further from the truth; *true* scientists are open minded, you have to be, science is not a list of dogma written in stone, it is the about the quest to understand our world including the weird things that happen in it. It is because of this that people are often scared of reporting sightings of any cryptid. Even one credible sighting is proof that there *could* be something there whether that be a new species, an oversized eel, sturgeon or something else entirely. The same newspapers have been left with egg on their faces the last few times they've published reports on Nessie's demise. When new sightings are made there is no reason to suppose it will be any different this time.

AQUATIC MONSTERS STUDY GROUP
www.cfz.org.uk

Ness be 'aving you

People were not always as sceptical of the Loch Ness Monster as they are today. In April, documents came to light that showed the strength of belief in, and just how seriously the local police took Nessie reports.

A few years after questions had been raised in parliament about protection of the monster for the first time (in 1934) William Fraser, the Chief Constable of Inverness-shire Police wrote a letter to the Under-Secretary of State in the Scottish office in 1938. In the letter Fraser said that the existence of a strange creature (he had written the word 'fish' but replaced this with the word 'creature' perhaps in view of some reports that claimed sightings of the creature on land) in Loch Ness was beyond doubt and raised his concerns that police powers may not extend far enough to protect the animal from people who wished to harm it. One example he cited was the case where a couple turned up at the loch with a harpoon and police were only able to verbally warn them against potentially harming the animal.

The full transcript of the letter is as follows:

INVERNESS-SHIRE CONSTABULARY
COUNTY CONSTABULARY HEADQUAR-
TERS
THE CASTLE
INVERNESS.
15th August, 1938

Sir,
The Loch Ness Monster.

I should like to refer you to your letter to me dated 21st Nov., 1933 (Ref. No. 36125/1) with which you enclosed a copy letter dated 13th Nov., 1933 received by you from Sir Murdoch Macdonald, M.P. for Inverness-shire.

In my reply to this correspondence, dated 23rd Nov., 1933, I indicated the only step which the Police could usefully take, was to warn the people resident in the neighbourhood and as many as possible of the visiting public, that the preservation of the Monster was desired.

It has now come to my notice, that a Mr. Peter Kent and Miss Marion Stirling, both of London, are determined to catch the Monster dead or alive.

Mr. Peter Kent visited Fort Augustus on Friday, 12th August, and was seen there by my Officer stationed at Fort Augustus, to whom he stated that he was having a special harpoon gun made and that he was to return with some twenty experienced men on the 22nd of August for the purpose of hunting the Monster down.

That there is some strange fish *creature* in Loch Ness seems now beyond doubt, but that the Police have any power to protect it is very doubtful. I have, however, caused Mr. Peter Kent to be warned of the desirability of having the creature left alone, but whether my warning will have the desired effect or not remains to be seen.
If you have any suggestion to make or can offer any guidance in the matter, I shall be grateful.

I am, Sir,
Your obedient servant,
William Fraser
Chief Constable.

Ersatz Escher?

Staying with the subject of Nessie, a previously unheard of painting, supposedly by the Dutch artist M.C. Escher, surfaced in the world's press at the end of March. The story goes that in 2005 an Italian traffic police officer, Raffaele De Feo, was clearing out the attic of his family home in Volturara when he found a strange picture of a faceless, naked dark figure playing a musical instrument while some sort of serpentine creature rises before him. Whereas De Feo and the press have been quick to insist that the creature in the picture is meant to be the Loch Ness Monster (presumably making the dark figure's instrument some very Scottish bagpipes) I have to say that it could just as easily be a drawing of a snake-charmer in India or a man eating a stick of rock while watching an emu eat an orange.

Regardless, De Feo did not take much notice of the strange picture until recently when he removed the frame and saw an inscription on the back of the charcoal drawing that read "With all my heart to a friendly remembrance" and M.C. Escher's signature.

The managing director of the company responsible for the copyright licensing of Escher's work, Mark Veldhusen, is unconvinced by the claims that the painting is a

lost work of the artist and has said that the company was never approached to verify the work. "This drawing is not made by M.C. Escher and the signature on the back does not belong to Mr. Escher".

This drawing was also checked by an art expert, Ms Dottore Anna Petrecchia. She apparently is a graphic art expert and is master in crime science and forensic science at the Court of Rome and said "I have no idea who is behind this scam."

"Since I have been handling originals for more than 25 years, the Foundation asks me to authenticate original prints, which I also do for Christie's and Sotheby's when there is doubt over a print.

"Anyone who has any common sense and has seen this picture can tell you that it is

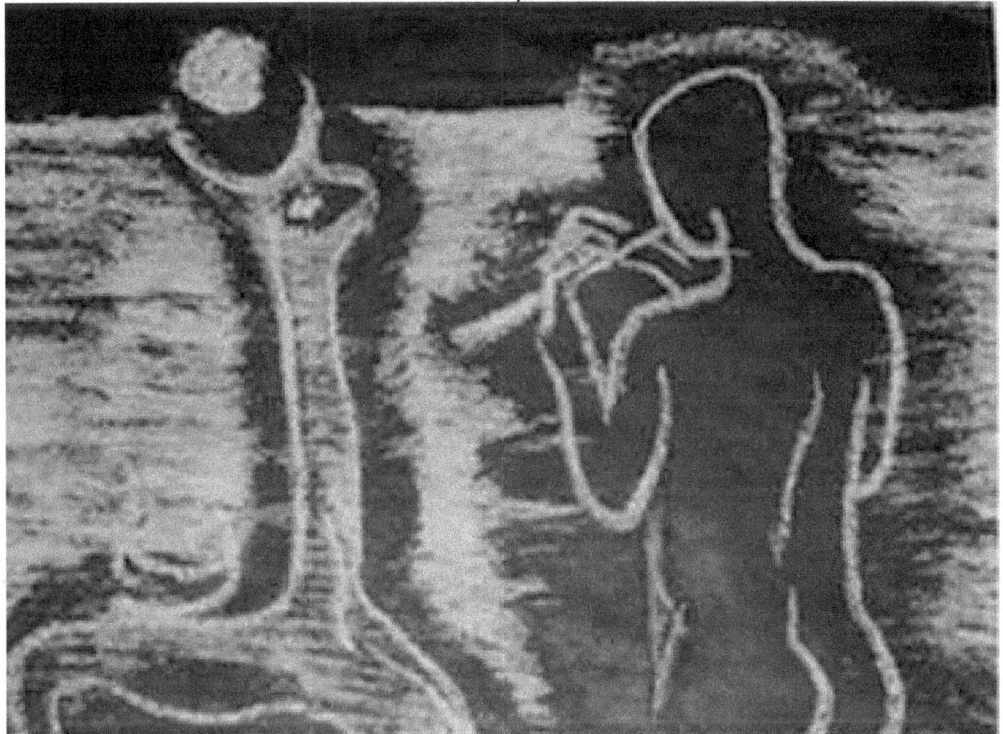

not by M.C. Escher.

"It is not his style, nor does the signature even remotely resemble his."

Veldhausen has shown the painting to M.C. Escher's son, George Escher, where upon…

"He burst out laughing when he saw it and said that his father would have stopped being a graphic artist if this was the best he could draw.

"Anyway, whatever they want with it, we don't know but I am pretty certain no auction house or respected museum wants to burn their fingers on this 'masterpiece'."

Even the most ardent supporter of the painting's alleged Escher origins would have to admit it looks far removed from Escher's more famous works which often involved highly intricate use of shapes and visual tricks with perspective. By comparison the 'lost' work looks very crude.

A Worthy Explanation?

In August 2009 'something' was filmed on Lake Worth creating a wide wake behind it, which led to the first reports of the muck monster as it is now known surfacing on the internet and in local news reports. One wake does not a monster make, as unusual wakes can be caused by a number of things, both manmade and natural, but Greg Reynolds, who filmed the wake seems convinced that the wake was made by an unusual animal.

Reynolds and a colleague, Don Serrano, were working for the not for profit organisation 'LagoonKeepers.org' cleaning debris from the lake when they saw the wake. Although one would expect that if they are out on the lake regularly the pair would be familiar with all wakes that they would regularly encounter,

Serrano described this wake as "Different, very different". The pair followed the wake, but whenever they got close it disappeared. The video was shown to Thomas Reinert, a marine biologist who works for the Florida Fish and Wildlife Conservation Commission for his analysis.

Reinert concluded that the 'creature' appeared to be moving in one direction and not breaking the surface, which in his view ruled out the wake having been caused by a dolphin, sea turtles, manatee or a large school of fish. Reinert also stated that if the wake had been caused by a shark a fin would have been visible. However, since Reinert saw the video the Florida Fish and Wildlife Commission have hit upon another explanation; the wake was caused by a manatee that had been injured by a ship's propeller. Scientists from the commission came up with the explanation after seeing groups of manatees huddled together in a warm channel of water by the Florida Power and Light Riviera Beach Power Plant during the winter and noticing a manatee with a three pronged tail amongst them.

The unusual tail of the manatee could explain the wake, although footage of the creature swimming would have to be obtained for comparison with the Lake Worth wake before any real conclusions could be made, but the manatee explanation wouldn't quite fit with the creature not surfacing during the video as Reinheart pointed out.

Call It Dave

John Kirk and the BC Scientific Cryptozoology Club found several large underwater objects while using fish-finder sonar on Cameron Lake, British Columbia, Canada last September. Kirk, the author of *In the*

massive object in the midst of various fish," said Kirk. "We were quite stunned that there was something that big in the lake, it was quite amazing."

The team made 4 more passes over the 74 feet deep object over the next twenty minutes to confirm that it wasn't just tightly shoaling fish. Kirk and the BC Scientific Cryptozoology Club are hoping to return to the lake in the summer for a longer study.

Domain of the Lake Monsters says that the sonar showed something that was certainly larger than a trout or any other lake fish but he wasn't ready to say exactly what it was. During the first pass of the lake the team made found two large objects at depths of 56 and 58 feet (depths of approximately 7m) and then on the second pass an object was found at a depth of 74 feet (22m).

"Something just went 'ping' on the alarm on the fish finder and we saw this absolutely

Only a Northern Pike...

There is a killer lurking beneath the waters of a Loughborough beauty spot. In March, ducks started to disappear from the surface of Stonebow Washlands, sucked down into the murky depths in the jaws of an unseen assassin. Locals have been warned to keep young children from pond dipping and not

to let their dogs swim in the water.

An un-named eyewitness, who told their story to the *Loughborough Echo* on the 31st of March said: "I was walking with my dog around the larger lake on Monday, and there were two mallards on the pond. "There was quite a commotion with the female making a lot of noise, and while I was looking she just disappeared.

"The male was just sitting there, so I walked round to get a better view and he just went down - all I could see was a ripple and three feathers on the water".

"It was quite upsetting." Mark Chapman, a wildlife development officer on the local borough council believes the killer to be a large Northern pike, and has said there are no plans to hunt the creature, known locally as 'Stonebow Jaws', down.

"Pike are a natural part of the ecology of our lakes, a native fish that have lived alongside wildfowl for thousands of years. This one would have to be a pretty big pike that's recently been put in. They aren't supposed to be there and people aren't supposed to fish at Stonebow Washlands.

We do take out all the introduced fish from time to time to stop people fishing, but we can only do that during the winter."

Crealy Good Fun?

In April the Cornish sea-monster Morgawr received an accolade reserved for only the most famous of cryptids when Crealy Great Adventure Park in Cornwall named their new roller coaster after the beast or zoo-form entity.

I can't say that the cars, fashioned to look like little dragons, bear too much of a resemblance to the famous 'Mary F.' photo but the ride looks like fun if roller-coasters are your bag and top marks should be given to Crealy for choosing to name the ride after a local cryptid rather than the somewhat unimaginative and generic 'Nessie' or 'Dragon' roller-coasters that seem to pervade theme parks these days.

The Strange Tale of the Cardiff Giant

The tale of the legendary Cardiff Giant is just about as weird, as surreal, and as convoluted as any tale can possibly get! And, without doubt, it was one of the most infamous and audacious hoaxes in American history. Essentially, the giant was nothing less than a 10-foot-tall purported "petrified man," said to have been uncovered on October 16, 1869 by workmen engaged in digging a well behind the barn of one William C. "Stub" Newell in Cardiff, New York.

In reality, however, the giant was nothing of the sort. It was actually the creation of a New York tobacconist named George Hull; an atheist, who decided to create the mighty-form after a heated argument with a fundamentalist minister – a certain Mr. Turk - about the passage in *Genesis 6:4* to the effect that giants once roamed the Earth. Hull's master-plan very quickly came to overwhelming fruition: he secretly hired a group of men to carve the enormous man out of a block of gypsum in Fort Dodge, Iowa, telling them it was intended to be used in the creation of a monument to Abraham Lincoln that would stand proudly in the heart of New York City. When work was complete, Hull shipped the block to Chicago, where he hired a German stone-cutter to further carve it into the likeness of a man – not forgetting, in the process, to swear him to absolute secrecy.

The ruse was a highly ingenious one: a whole variety of stains and acids were used to make the giant appear both ancient and weathered. In addition, the giant's surface was beaten with steel knitting-needles embedded in a board. The purpose: to simulate pores on the skin. If nothing else, Hull had carefully and skillfully thought out his grand-plan. Then, in November 1868, Hull transported the giant by rail to the farm of his cousin, William Newell. No less than $2,600 was spent on the hoax in total – which was a sizeable amount of money, indeed, way back in the 1860s.

Almost twelve months later, Newell hired Gideon Emmons and Henry Nichols, ostensibly to dig a well, and on October 16, 1869, lo and behold they "found" the Cardiff Giant. One of the men reportedly exclaimed, in excited and exaggerated tones: "I declare: some old Indian has been buried here!"

But that was only the start of the matter: Newell quickly set up a tent over the giant and charged 25-cents for anyone and everyone who wanted to see it. Two days later, very pleased by the huge number of people who turned out to view the Cardiff Giant, he increased the price to 50 cents. Enterprise was truly the name of the game.

Archaeological scholars quickly pronounced the giant nothing more than a fake; while a number of geologists noticed there was no logical reason for digging a

well in the exact spot the giant had been found. And Yale palaeontologist Othniel C. Marsh came right to the point, famously declaring the Cardiff Giant a "most decided humbug". There were, however, some gullible Christian fundamentalists and preachers who defended its legitimacy.

Ultimately, Hull sold his part-interest for the very impressive sum of $37,500 to a syndicate of five men headed by one David Hannum. They, then, clandestinely moved the giant form to Syracuse, New York for exhibition. Unsurprisingly, the giant drew such massive crowds that the famous showman P.T. Barnum offered $60,000 for a three-month lease of the giant. When the syndicate flatly turned him down, however, the always-resourceful

and industrious Barnum hired a man to create a plaster replica – which quickly went on display in New York, amid claims that this was the real thing, and that the Cardiff Giant was the hoax!

As newspaper journalists gleefully reported on Barnum's version of the story, David Hannum was quoted as saying, "There's a sucker born every minute;" in reference to spectators paying to see Barnum's giant. Over time, the quotation was misattributed to P.T. Barnum himself. Hannum then tried to sue Barnum. In somewhat humorous fashion, however, the judge told Hannum to get his giant to swear on his own genuineness in court if he wanted an injunction in his favor. That was a tough one to achieve.

But still matters were not over: on December 10, Hull confessed the truth to the press. Then, on February 2, 1870 both giants were revealed as fakes in court, and the judge ruled that Barnum could not be sued for calling a fake giant a fake. And that was the end of the lawsuit, not surprisingly. And the events stirred up something else too: they encouraged others to come

forward with their very own versions of the Cardiff Giant. As evidence of this, in 1876 the "Solid Muldoon" surfaced out of Beulah, Colorado and was exhibited at 50 cents a ticket. There was also a rumor going around that Barnum had offered to buy it for $20,000. It was, needless to say, a fake – and possibly one that George Hull himself had a hand in.

One year later, in 1877, the owner of Taughannock House hotel on Cayuga Lake, New York, hired his own merry band of men to create a fake petrified man, and who carefully placed it precisely where the workers that were expanding the hotel would eventually find it. Once again, publicity and public interest were impressive. But, it was still a hoax. Then, in 1892, a certain Jefferson "Soapy" Smith, the de facto ruler of the town of Creede, Colorado, bought a petrified man – "McGinty," as he became known - for $3,000 and exhibited it for 10 cents a look. Interestingly, this giant was actually real. That's right:

a human-body, deliberately injected with chemicals for preservation. Soapy enthusiastically displayed McGinty from 1892 to 1895 throughout Colorado and the northwest United States. Seven years on – 1899, to be precise - a petrified man was said to have been found in Fort Benton, Montana. The body was supposedly identified as that of U.S. Civil War General Thomas Francis Meagher. Meagher had drowned in the Missouri River two years previously. The petrified man was transported to New York for exhibition; but, needless to say, it was not the general, at all. The Cardiff Giant – which started all the fuss – continued to surface from time to time. However, its place in the limelight was clearly waning. In 1901 the giant appeared on display at the Pan-American Exposition, but failed to generate any significant attention or publicity.

Then, some years later, an Iowa-based pub-

lisher purchased it – for use as a coffee-table, no less! Seemingly eventually growing tired of the giant, in 1947 the man sold it to the Farmers' Museum in Cooperstown, New York, where it is still on display. And there's another very good reason why the controversy of the Cardiff Giant refuses to roll over and die.

In January 2010, I gave a lecture on cryptozoology at Cooperstown – as part of the *Ghosts of Cooperstown* event that included presentations from the people behind the History Channel's *Ghost Hunters* series. After the lecture, a local woman named Sally came up to me and swore that her father had seen the Cardiff Giant striding through the woods of Cooperstown late one winter's night in 2007. I asked Sally if she was joking. She was not. Her father, Sally said, had been driving home at around 11.00 p.m. one Friday night – after visiting friends in Albany, NY.

As Sally's father approached one particular stretch of road enveloped by trees, he was shocked to the core by the sight of the Cardiff Giant looming out of the woods and striding across the road in several mighty steps. Not surprisingly, he hit the brakes. Sally told me both she and her father, as locals, had been to see the Cardiff Giant on display at the local museum on several occasions over the years. They also knew full well that it was nothing more than a century-old hoax. So, how could a hoaxed creation be seen wandering the chilled woods of Cooperstown in 2007? Sally's opinion was that the Cardiff Giant that her father saw was not the same entity that currently rests in the Farmer's Museum. Rather, she felt, he had been blessed with a sighting of a thought-form – a Tulpa – that had been conjured into existence by the sheer unconscious will of those who wished to believe it was real. It was a mind-monster, in other words – one that cannot exist unless

people believe in it. Could it really be the case that a mind-originated Cardiff Giant haunts the darkened parts of Cooperstown, one destined to roam the neighborhood by night, until such a time that a lack of belief in its existence dooms it to inevitable annihilation?

It was as good a theory as any, I suggested. And, as someone who has dug deeply into the world of thought-forms and Tulpas, I reasoned it made a great deal of sense, too. The saga of the Cardiff Giant, I suspect, is far from over…

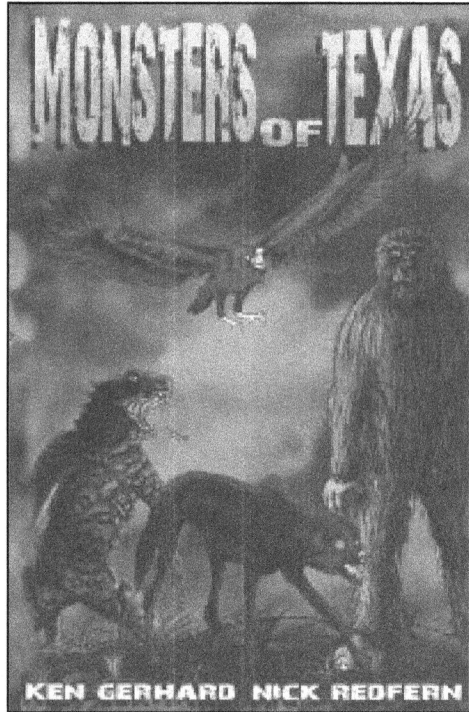

Nick Redfern is the author of many books on the paranormal, including *Man Monkey* and *Monsters of Texas* (co written with Ken Gerhard)

MYSTERY CATS DIARY

Suffolk ABC report - Kirton, near Trimley St. Martin

The Coastal Advertiser, a weekly freesheet for the Suffolk Coastal district - had a story in the March 12 edition - "Claws back on the prowl"

"Claws" is apparently the already established name for "the county's very own big cat".

The date of the sighting isn't given, but it was presumably not much more than a week before 12 March. Taxi driver Paul Smy, who runs Paul's Travel, spotted a "large cat-like animal crossing a field at Kirton as he pulled up at the junction of Park Lane and Bucklesham Road." (this is a fork in the road at the north end of the village of Kirton is two miles north of the A14 Ipswich Felixstowe Rd, near Trimley St Martin, which is on the outskirts of Felixstowe.)

Mr Smy said of the cat "it was a quarter of a mile away, and yet it looked really big." Mr Smy stopped his car in the road to watch the animal and was then "beeped" by a woman in a car behind him "on the school run" (which would have made it an early morning sighting). Mr Smy then got out and pointed out the big cat to the woman "and we both watched it together as it made its way across the field. It was just moving slowly without a care in the

world." Mr Smy said he'd lived in Africa for several years where he had "seen lions and tigers and leopards at close quarters. I would say this animal was about the size of a leopard" but he reportedly "believed it to be a panther." He added that there were sheep in some of the fields locally.

The news article mentioned that "Claws" had been seen "creeping through the country's woodlands on the hunt for prey" over the years, starting from 1996, citing an incident in the city of Ipswich's Foxhall Lane in that year, in which "startled residents watched as a panther-like beast weaved through the undergrowth." The article added that the (Ipswich) *Evening Star* newspaper had been "charting sightings of the big beast for the past 15 years."

Big Cats in Britain's virtual map of sightings for recent years shows reported sightings are less dense in Suffolk than in the rest of the UK. This may be because the county is relatively sparsely populated. You do, after all, need people to make sightings. Sightings on the BCIB map are more dense around suburban Ipswich and Newmarket, where the human population is bigger. There are an awful lot of places in rural Suffolk where big cats could hide, especially with current initiatives like the Suffolk Broads, in which Suffolk Wildlife

Trust are buying up plots of land with the aim of creating a north-south 'corridor' of fenland nature reserve across the country.

The article doesn't mention the colour of the big cat spotted, but an earlier wave of sightings - peaking around 1999 and concentrated in the countryside around Bungay in the north of the county - resulted in remarkably consistent descriptions of a black cat with a very long, straight tail and slightly bigger than a labrador. (The animal in the 1999 wave would seem from descriptions to be smaller than the animal in Mr Smy's description.) There was also a European lynx shot near Beccles in 1991, kept in a freezer for several years by the farmer that shot it, terrified as he was at what Defra (or its contemporary equivalent) would do to him if he reported having shot it. (*East Anglian Daily Press*, March 16 2006, S'uffolk *Tales of Mystery & Murder*, Mark Mower, Countryside Books, Newbury, Berks.)

This could be another reason for the comparative scarcity of reports.

More and more Suffolk farmers are cooperating on environmental initiatives - often receiving financial incentives in lieu of the old subsidy programmes - but there may be farmers who for whatever reason don't feel it's in their interests to cooperate on the latest environmental initiatives and are quietly shooting big cats and keeping it quiet. The lynx was shot in 1991 after 15 sheep had been killed locally within a week, after all.
MATT SALUSBURY

2009 BIG CAT ROUND-UP

By Neil Arnold

I've rarely written up any official summary of sightings and evidence, except for the occasional piece thrown to a local rag in order to receive more reports.

However, on this occasion I'd like to share with you several stories pertaining to my research throughout 2009. I'll certainly try to refrain from the tiresome listings which continually pepper just about every research group's 'official' press release, which usually compiles of inaccurate regional reports which read as boring "It was a big, black Labrador-sized cat and it crossed the road" statements.

2009 was a reasonably quiet year throughout the south-east. Was this because the cats that once inhabited the area had moved on? Had they died? No. It was simply down to the fact that I'd spent most of the year writing instead of getting out into the field and bombarding the press with appeals to the public. On average I receive

THE CENTRE FOR FORTEAN ZOOLOGY

MYSTERY CATS STUDY GROUP

around 250 eye-witness reports a year (this figure comes after sorting out the misinterpretations, hoaxes et al), and this had remained consistent since 1998. 2009 was down to about 150, but produced some excellent cases.

All the usual rural areas continued to produce sightings of puma, black leopard and lynx – i.e. Maidstone, Ashford, Canterbury, Gravesend – and of course Sussex, but the most intriguing reports emerged from London, where on several occasions throughout the year a black leopard was seen around Sydenham, Norwood, and Crystal Palace. Tragically, the local papers dubbed this creature the 'Palace puma' despite the area being bereft of a puma sighting. It all came to a head however, in December when jogger Roger Fleming was running through a wood in West Dulwich, with his dog, when he was suddenly chased by a large cat. Fearing for his life after a few metres, Roger was about to submit to the beast, when he turned and found no trace of the cat (I can hear the crackpot researchers already shouting, "It's because big cats are paranormal!!"). Had this been the same creature, dubbed the 'beast of Sydenham', which was alleged to have attacked a man in his back garden in 2005? Well, it's unlikely the attack in '05 ever occurred but Roger was positive that he'd been pursued by a large, dark-coloured cat which may well have been after his dog. It was possibly the same animal that had eaten a domestic cat in Sydenham's Southend Park in September. The most controversial sighting around London took place in March at Peckham Rye when a dog-walker claimed to have sighted a pair of cats with 'leopard-like' patching on their coats. I immediately dismissed the possibility that these animals were leopards with normal pelage which caused uproar in the 'big cat' community who responded by stating, in as many words, that I was wrong to state that there were no sightings of 'normal' leopards. Judg-

ing by research I've conducted since I was fourteen, there are no 'normal' leopards roaming the south-east, and I've based this on lack of consistency in such reports, if reports exist at all. Oh, how Sergeant Leopard's Phoney Old Fart's Pub Clan despised my claim.

Sceptics continued to dismiss sightings of a 'big cat' around London, without once leaving their armchair to actually look at some of the more leafier areas around the capital which could easily hide an elusive, nocturnal predator.

In February 2009 as frost hardened the Kentish soil, Paul Knowles, whilst golfing at Southern Valley Golf Club, Gravesend, found a set of mysterious prints crossing a bunker. The prints appeared to have been left by a creature measuring around four-feet, and walking on all fours. Although the local press were quick to bring the 'beast of Bluewater' to mind, Howlett's Zoo in east Kent dismissed the prints as belonging to a dog. The main problem with this conclusion was the fact that the prints showed a long heel, similar to prints made by a wolverine, or another mustelid, the fisher, a medium-sized mammal, although the prints were almost as long as the size ten boot print of the witness!

Sceptics argue that credible witnesses never come across evidence of large, exotic cats in the wild. This is of course complete rubbish. In December '09, Mr Wilde, who belongs to the Metropolitan Police - whose training centre is based at Gravesend - found a set of footprints in the snow that he was 100% sure belonged to a large cat. He photographed the prints which were found on remote land where dog-walkers do not venture, and they were sent to me. The

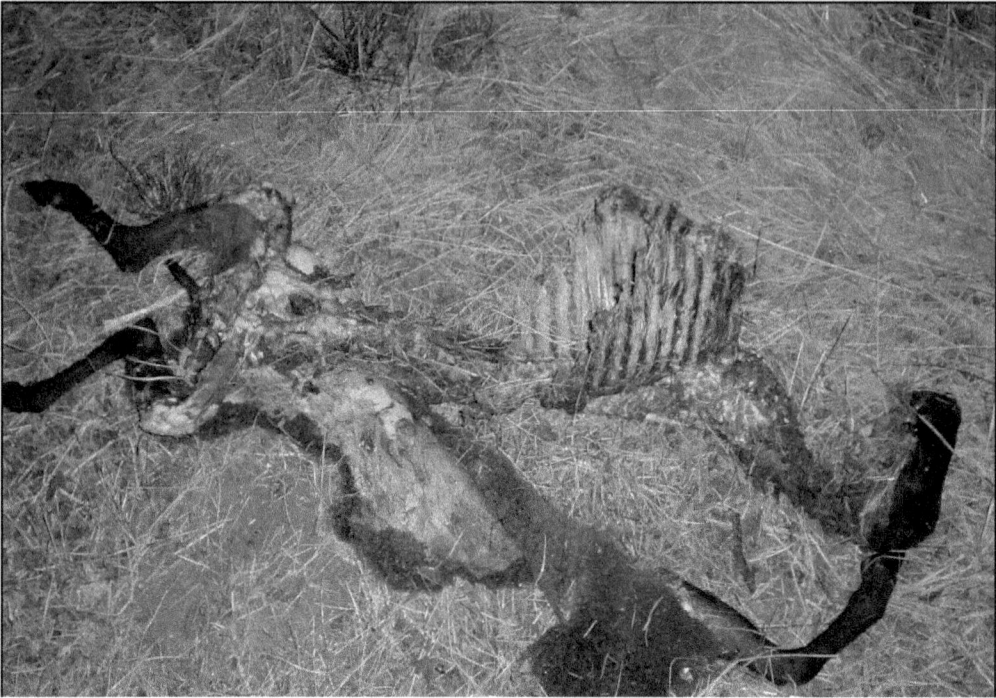

ABOVE: Bredhurst sheep kill August 2009
BELOW: Domestic cat partially eaten, Denham 2009

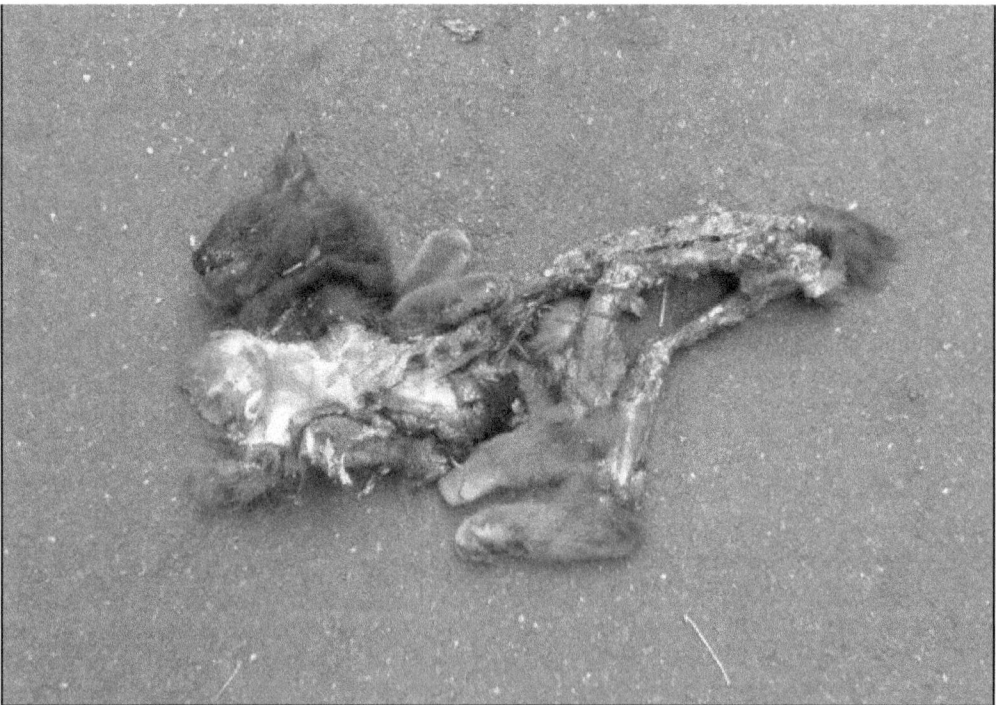

ABOVE: Goudhurst mystery cat July 2009
BELOW: Lamb carcass Trottiscliffe May 2009

prints clearly belonged to a large animal which, as it walked tended to place its rear paw over its front paw prints. The prints measured around five-inches and were asymmetrical and stretched for a kilometre. Similar prints were also found in an isolated spot in Sittingbourne, where an animal had emerged from the woods and walked over the roof of a parked vehicle.

Other evidence emerged in 2009 of kills. An eaten deer was found by an alarmed mother whilst walking with her children through an area known as Devil's Dyke in Brighton (Brighton would figure heavily in regards to cat sightings, the most impressive concerning pro' golfer Ian Campbell who spotted a black leopard whilst practising on the main golf course in the region). Several sheep attacks took place in 2009, the most conclusive on the outskirts of Medway, and blamed on the so-called 'beast of Blue Bell Hill'. I visited the land of the farmer in question who commented that in forty years they'd never lost any sheep until one such unfortunate individual was found completely stripped clean on the fringe of heavy woodland.

In 2009 I investigated the case of the Rochester 'rabbit ripper' – an elusive predator keen on eating local bunnies and dumping them in back gardens, along with a few cleanly stripped domestic cats. I also investigated the 'beast of Brentwood' (Essex) which on several occasions had stalked London cabbie Mr Toulson as he walked through a park with his dogs. There was also the summer which brought with it some excellent photo's of a Bengal cat, of interesting pelage in which zoologist and friend Karl Shuker took great interest. I also went on the track of an African hunting dog, but there were a few disappointments too, namely the poor

little domestic cat caught on film at Thanet which the local newspaper swore blind was a 'panther', and the various lumps of alleged 'big cat' pooh I got sent through the post for analysis!

Despite the ups and downs, being a full-time researcher has enabled me to speak to thousands of people throughout the year, from Devon to Essex, and from Surrey to Sussex, to enable the public to understand these animals more, separating the facts from the fiction. It is a privilege to write and speak about these animals, because if it wasn't for such individuals roaming the wilds, I wouldn't be the person I am today.

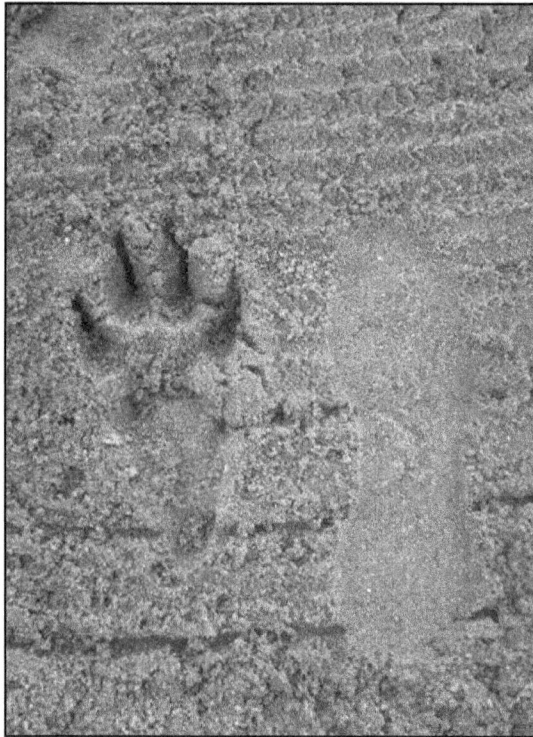

Pawprint found at Southfleet during 2009

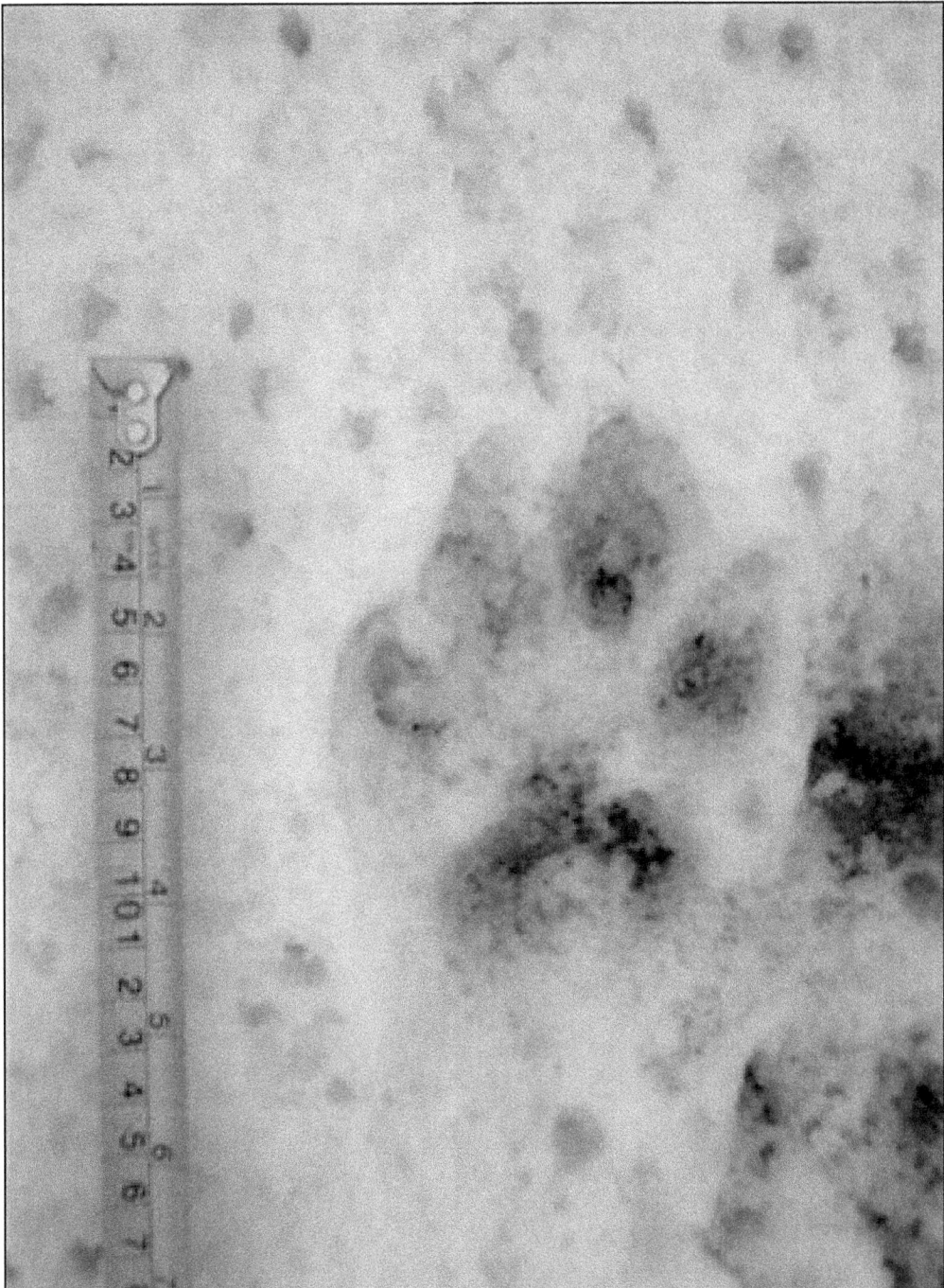

Gravesend pawprint Dec 2009

THE CFZ DOWNUNDER

The Yowie – A Window Into Australia's Bigfoot Enigma

by Rebecca Lang and Michael Williams

"*It was like an elephant on two legs wearing size 20 boots.*" That's how Blue Mountains resident Neil Frost would later describe the ape-like creature he encountered in thick bush near his backyard in February 1993 to local police.

The schoolteacher was the then-father of a newborn son, and loathe to wake up his youngest child with a midnight trip to the toilet. He chose instead to sneak out the back door and relieve himself in the bush-fringed yard. It was as he finished his business that a large shape loomed out of the shadows, surprising him before taking flight.

"*I'd just finished and taken one or two steps and this thing got up out of the bush and just took off,*" Neil recalled.

"*It was unbelievable. It sounded big, like it weighed...I don't know...maybe 130 kilos. I didn't get a good look at it.*"

The chase

A somewhat shaken Neil bolted next door to alert neighbour Ian `Lizard' Price, a herpetologist and former bikie gang member. A generally fearless character, Ian was surprised to find his normally "quiet and boring" neighbour banging on his bedroom window in the middle of the night.

"*He was in a state of really extreme agitation,*" Ian recalled.

"*So I said 'Let's do it'. After all, the most it could do was just kill us!*"

Together, the pair pursued the shape into the bush, but the creature eluded them – always just one step ahead of the two men – and began what was to become a bizarre, ongoing ritual of hide and seek. Reporting the matter to local police proved a fruitless exercise – the two men later found out the local constables had carried out a door knock to find out if either of the pair were known troublemakers or "doing drugs".

So night after night, the two men would chase the creature - which would announce its presence by noisily thumping the ground – watching it flit between trees and across fire trails just beyond their reach.

In their search for the elusive apeman, Neil and Ian also located what appeared to be full body imprints in a bed of reeds near their homes and numerous footprints – the latter Neil and Ian initially disregarded as some kind of hoax.

"*I'm reasonably well-versed with all of the indigenous wildlife in this country, to the extent I could identify any animal that you bring up, if not species at least genus,*" Ian said. "*But this was strange. It was 7ft tall, two of me in build. I weigh 100kg, so we're looking at a creature at least three times my body mass. It was big. It ran sort of like a person, but not quite. There*

THE CENTRE FOR FORTEAN ZOOLOGY

www.cfz.org.uk

52

was something odd in its gait.

"It had a testicle-tightening growl, creating a real predator-prey type reaction. People say you shit yourself when you're really scared, but you don't. Your bum sucks in.

"And it could out-run us, out-see us and avoid barbed wire fences. It was a big hairy beastie."

Despite the frightening spectacle, Ian retained his sense of humour about the risks involved with chasing something so large and, presumably, dangerous.

"I guess I'd be famous posthumously," Ian chuckles. *"I said to Neil at the time: 'If it rips my arms off and stuffs them down a hole, name it after me."*

On another occasion, Neil was lucky enough to get within a few feet of the creature after a visitor was adamant *"something had followed him down the driveway".*

"I held up my torch and shone it in front of me and this thing just stood up, leaned into me and roared into my face," Neil recalled. *"Then it took off through the bush. It was pretty terrifying, but what I remember of it was that it had a large head and red eyes.*

It roared again, and there were dogs barking right across the valley. I remember people were coming out of their houses to tell their dogs to be quiet."

More sightings

The incidents in 1993 set the scene for multiple sightings of the creature near their homes, situated in one of the many Blue Mountains townships along the Great Western Highway. The road snakes its way through a massive world heritage-listed wilderness area 1.03 million hectares in size, consisting mostly of forested landscape. Among Fortean researchers the area has acquired a reputation for "high strangeness" due to the proliferation of UFO, big cat and yowie encounters and sightings, not to mention the obligatory haunted homes dating back to the early days of colonial settlement. *"You've got all this land mass that's basically unexplored,"* Ian points out.

"Look at the Grose River Valley, no one goes there. Look at the (recently discovered) Wollemi Pine. It's a 100ft tall tree. It can't run around and hide. If you've got a 7ft tall 300 pound creature that can hide..."

The Wollemi pine (*Wollemia nobilis*), the so-called "dinosaur tree" thought to be long extinct, was rediscovered in 1994 by a park ranger exploring a deep, narrow canyon in the heart of the Blue Mountains.

At least 63 of Neil and Ian's neighbours have also heard or seen the bipedal creature near their own homes or in the dense, largely unexplored bushland that surrounds the townships - both before and since that day.

One, a nine-year-old boy living two doors away from Neil, was playing with fire one day and ran through the bush with a burning stick in his hand.

"As you can imagine, that's not a very smart thing to be doing around here," Neil said, referring to the frequency of devastating bush fires in the area. *"He ran smack-bang into him, literally into this creature – it looked down at him, he looked up at it and they both ran in the opposite direction to each other completely horrified."*

"He gave a beautiful description of the creature, describing the skin folds on its face as leathery and black, with thick hair on the outside - grey with white flashes through it."

Monsters on the loose

On the surface, one would have thought monster-sized primates frequenting heavily populated areas in the Australian bush would - by now - surely have been discovered, identified and named by science.

Unlike the United States, which has larger wildlife like bears, Australia has no animal that could be mistaken for the yowie or yahoo, the Aboriginal names given to the bigfoot-like creature Neil and Ian chased. Nor does Australia boast any non-human primate lines such as monkeys or great apes. While anecdotal reports are plentiful, there is a distinct lack of credible photographic, video or audio evidence of the creature. Not one whole or partial skeleton has ever been found of the creature. Neil's own efforts to collect samples for DNA-testing have also met a dead end, but he has vowed to continue his quest to prove something physical is out there.

Bizarre plaster casts and photos of yowie footprints that have been taken in the area show giant two, three, four and sometimes five-toed footprints and do not look as though they could sustain a large animal's locomotion, let alone stabilise the creature. No known species displays the same gross morphological variation that the yowie's feet seem to do. They don't even look remotely primate-like.

Neither man favours the other obvious explanation for a creature that does not feature in Australia's zoological profile, can elude capture, move with lightning speed and is often reported as having glowing red eyes – that is, a creature of paranormal origin.

Tapping the Dreamtime
Struggling to find out more about his encounters, Neil eventually approached local Aborigines from the Burragorang tribe who were, initially, wary of discussing the yowie with a white man. Eventually, after Neil imparted his story, they told him what they knew of the creature from their own personal experiences and as it featured in their Dreamtime stories.

Described as tall, powerfully built and with arms that tapered into claws, the yowie was reputed to be a man-eater. It was also usually accompanied by an overpowering stench, an element that often crops up in many modern-day yowie sightings. But some of the Burragorang people's answers merely added to the enigma for the schoolteacher.

"The Aborigines told me that they live in the trees, but I think that's just an illusion that they have. In fact what I think they meant is they just stand behind trees," Neil said. *"There was also an Aboriginal story that yowies have their feet on*

backwards. I don't believe that they do, I think what they meant, my interpretation, is that they're just f***ing hard to track."

The reference to backward feet is a common one in faerie lore, and regularly crops up in bigfoot/ yeti stories from around the world. It also lends the creature a somewhat otherworldly air. But ask Neil his opinion on the phenomena, and he's adamant: *"It's definitely a real creature. I don't buy into any of this paranormal stuff."*

Perhaps Paranormal?
One researcher who does favour a possible paranormal explanation is veteran Australian 'monster hunter' Tony Healy.

Healy, who along with colleague Paul Cropper* is writing what will be the most comprehensive book ever put together on the subject of the yowie, believes some accounts point to the creature *"being something other than an anthropological or zoological mystery, that is, something other than flesh and blood."*

"If we reject everything about the yowie that smacks of the paranormal we'd have to sweep 20 per cent of the accumulated data under the carpet," Healy told the Sydney 2001 *Myths and Monsters Cryptozoology Conference***.

"After 25 years on the trail I really suspect that the American Indians, the Australian Aborigines and some of the whacked-out American researchers are right - that is, that we're dealing with shape-shifting phantoms here that will probably remain beyond human comprehension."

How else to explain the vanishing acts, glowing red eyes (in one encounter described as "the size of tennis balls"), superhuman feats of speed and the ability to invoke the so-called "nameless dread" – the inexplicable primal fear that "turns your guts to water", experienced by so many witnesses?

"The dread", as some self-styled yowie hunters refer to it, has stopped many individuals in their tracks and inspired a hasty retreat to the safety of cars and civilisation. It is not unlike the sort of temporary terrifying paralysis that strikes the prey of big cats. What kind of creature could engender that kind of fear?

Lions and tigers and... yowies...oh my!

Jerry and Sue O'Connor are neighbours of Neil and Ian, and they firmly believe that the yowie is of paranormal origin. They favour the term "nature spirit" - and given some of the examples of paranormal attributes outlined thus far - they may yet be the closest to the mark in truly describing the mysterious creature. Their first encounter with the yowie was in November 1999. Sue was out painting on the couple's back verandah late one night when the silence was broken by a spine-tingling scream.

"It was about 12am and I was just doing some painting, when suddenly there was this sound," she said. *"I don't know what it sounded like, like something out of this world - like a cross between a lion, a bear and a monkey. It sort of growled-roared. It freaked me out; sent shivers up my spine. So I ran inside and called for Jerry."* Initially sceptical of his wife's claims, Jerry walked out to the back verandah, explaining to Sue as he went that the noise, being springtime, was probably just wallabies or wombats making mating sounds.

"Then this thing absolutely roared," Jerry recalled. *"It was alien, and it sounded like nothing I've ever heard in my life. We didn't have a torch so we couldn't really see anything. But the feeling that came from it was like a physical barrier, like I was a threat to it. This wall of fear hit me and the hair stood up on my arms and the back of my neck. I just dragged Sue inside and shut the door."*

"There was nothing that I was taught about in the Aussie bush that sounded like this. I never wanted to hear anything like it again, it was that frightening."
The couple, both in ill health, had moved to the area two years prior "for a quieter life" following

Jerry's discharge from the Navy. But that encounter shattered more than the peace and quiet of one evening – like a ripple effect in a pond, it was the start of something much bigger.

From myth to reality

"I thought that yowies were just something we had invented because the Americans had Bigfoot," Jerry said. *"That incident was the first inkling that we had that there was something bizarre living in the bush around here."*

Like Neil and Ian, the O'Connor's encounter set the scene for a number of bizarre incidents around their home, leading them to become reluctant witnesses to a creature they had relegated to storybooks. Not knowing Neil at the time, the O'Connors got in touch with Queensland-based Dean Harrison, who runs *Australian Hominid Research****, a group dedicated to researching the yowie phenomenon.

Dean put the pair, who lived just streets away from each other, in touch with one another, passing on the story of a third man who had spotted something large, hairy and hominid on a road near the O'Connor's home – the same night they had experienced the spine-tingling roar.

"This bloke Brad Croft was driving along the road, which is about 500m away our house, when he saw this large, 8-9ft hairy man cross the road and go down into the gully," Jerry said. *"He drove straight home and got Neil, who lived in the same street as him, and within 10 minutes they were back at the spot walking around. As they got down near the creek they heard it crashing around and running up a track, which, as it turns out, led straight to our backyard."*

Night visitors

On another occasion soon after, Sue was sleeping on her bed with several of her six cats when she was woken up by the sound of their collective growling. *"Their hackles were up. I looked up to see why they were growling - what they were looking at - and just caught a flash of a shape in the window, about eight foot tall,"* Sue recalled.

Prior to that, the couple had regularly been woken up at night by the sound of their external fuse box cover being opened, and banging on the side of the house. *"The fuse box door would slam, but there were no kids living in the area at the time, and hardly any neighbours,"* Jerry said. *"We thought that it might have been yobbos [pranksters], but something was happening every week."* A short time later Jerry, too, caught sight of their hairy visitor at the bedroom window, firming his resolve finally that something was afoot.

"God knows how long this thing had been looking into our window for," Jerry said. *"I woke up one night and looked at the window and I saw this thing – it was like a silhouette of a human head, with a messed-up lump of hair and a nose like a human had - not like an ape with pushed-back nostrils. It had these massive bloody shoulders, like a gridiron player wearing shoulder pads. I was convinced then. I wasn't exactly a sceptic before, but this convinced me completely."*

As the frequency of the sightings increased, the couple began to have bizarre dreams about the creature, often waking in time to catch sight of a hairy face peering through the window. *"I noticed we started having these nightmares. I would wake up and say to Sue, 'I've just had this frightening dream of a yowie' and she would say 'Really? So have I'."*

Faerie folk?

Footprints were also found on the couple's property after some of these visitations, which Jerry photographed and cast. The couple, who regularly bushwalk, have also spotted strange footprints along the fire trails and tracks that criss-cross the area, leading down into gullies, waterfalls and caves. Despite these monster-like physical traces, they remain convinced the creature is more faerie-like than fierce physical beast.

Aboriginal lore does support the existence of smaller hairy men in Australia, the yuuri (pronounced 'yawri', not unlike yowie) or 'brown jack', which fulfill a similar role to that of European elves and leprechauns – they guard certain places, grant favours and play tricks on people.

The mystery lingers on

While the ongoing experiences of the Frosts, Prices and O'Connors - which we have only touched on very briefly - are confined to the Blue Mountains region, there have been many yowie sightings all over Australia, predominantly along the east coast in mountainous and forested areas.

But despite the prolific number of sightings, the occasional compelling plaster cast of a footprint and a rich oral history about these creatures that predates white settlement by thousands of years, there is, as yet, no concrete proof that a hitherto unknown species of apeman walks the Australian continent. The elusiveness of the creature, if indeed that is what it is, continues to puzzle and frustrate those who have encountered it, and those, like the authors, who pursue proof of its existence.

It may one day be shown that the paranormal and physical aspects of the creature are indeed the same sides of the one coin. In the meantime, there can be no satisfaction. All we can do is flip it and see.

References

*Healy and Cropper also authored the Australian cryptozoological classic Out Of The Shadows, Pan Macmillan, 1994.

**The 2001 Myths and Monsters Conference was the first major gathering of cryptozoologists ever held in Australia. The conference, organised by Ruby Lang and Paul Cropper, featured a host of academics, amateurs and enthusiasts.

***Dean Harrison's AYR website www.yowiehunters.com has facilitated the collection of many hominid sightings across Australia.

watcher of the skies

CORINNA DOWNES

Bernard Heuvelmens said that "*cryptozoology is the study of unexpected animals*". We are adding this section as a new feature to *Animals & Men* and in this issue are some of the rarer birds that have been seen in the UK since the last issue of the journal was published. Over the coming issues we hope to be able to list birds that are very rare and also some that may be particularly unexpected.

The British list comprises all those bird species which have occurred in a wild state in Great Britain, and in general the avifauna of Britain is, of course, similar to that of Europe, although with fewer breeding species.

There are 587 species of birds on the British list as of 22 July 2009, with the latest additions being the Pacific Diver (*Gavia pacifica*), Yellow-nosed Albatross (*Thalassarche chlororhynchos*), Glaucous-winged Gull (*Larus glaucescens*) and Asian Brown Flycatcher (*Muscicapa dauurica*).

Glossary

British Ornithologist Union (BOU) list Category A: a species that has been recorded in an apparently natural state at least once since 1 January 1950

AEWA - Agreement on the Conservation of African-Eurasian Migratory Waterbirds

There have been quite a few rare vagrant bird sightings to our shores since Christmas, and it has been almost impossible to include them all in these pages. However, the most interesting has to be the first nesting pair of purple herons in Kent. Since compiling the collection in this edition, the recent news that this pair have successfully hatched their brood on the RSPB's Dungeness Nature Reserve is both exciting and encouraging, and two flew the nest in early August.

We can but hope that this means that purple herons will become a more common sight in this country. Let us keep our fingers crossed that the parents manage to rear their young successfully.

The following birds of note have been spotted as below:

Ardeidae

May - Dungeness, Kent
Purple Heron (*Ardea purpurea*)

A pair of purple herons was found nesting in Dungeness, Kent during May. This is the first time ever that this bird has nested here and it is hoped that their nest will produce first fledglings.

However, this newest breeding bird species is under threat from an airport development in Kent. The purple heron usually breeds in Southern Europe and only visits in small numbers each year to the UK – usually as individuals.

Dr Mark Avery, RSPB Conservation Director, said: "The RSPB has a century-long heritage of protecting birds in Kent. Our protection scheme at the site provides immediate security for the birds, but the potential development of an airport on the peninsula casts a large shadow over the future of this magnificent site and its wildlife. Purple herons are high up on the list of birds that we expect to see setting up home in southern Britain as the changing climate pushes them further north. This highlights the importance of wildlife havens like Dungeness in providing space for species displaced by global warming."

Experts expect the number of purple herons in Britain to increase in the years to come, although they have struggled in Europe over the last few decades. Both cattle egrets and spoonbills have bred for the first time here and great egrets are being seen more regularly, and may well become the next colonist.

The RSPB has set up a round-the-clock Spe-cies Protection Scheme to protect what could be the first successful nesting ever recorded in the Britain and Kent police are assisting.

PC Michael Laidlow, Environmental Crime Coordinator at Kent Police, said: "We would remind any members of the public that any attempt to intentionally damage an occupied nest or remove eggs constitutes a criminal offence. This is a popular area for birdwatchers and the local community so I would urge anyone in the area who sees any suspicious activity to call the RSPB or Kent Police." Dr Avery added: "At the moment there is little for visitors to see as the birds are sitting tight on the nest. The area is a great place to visit and is teeming with spring wildlife – but please don't come hoping to see the herons as you are likely to be disappointed. If the eggs hatch successfully then we hope to set up a viewing station at a safe distance away from which the public will be able to catch a glimpse of this historic wildlife moment."

Meanwhile the RSPB is calling on the government to call in the planning applications for proposed expansion at nearby Lydd Airport. The local authority Shepway District Council controversially consented the applications in the face of a recommendation to refuse, given on environmental grounds by the council's own planning officials. So far over 10,000 representations have been made to the Government Office of the South East to ensure this decision is scrutinised in a full public inquiry.

Procellariidae

June – Lundy Island, Devon
Little Shearwater (*Puffinus baroli*)
BOU/IRBC Category: **A**
RBA Status: **Extremely Rare Vagrant**

At going to press, a male little shearwater

of the form baroli has been present for two nights, but has only been heard singing from its chosen burrow but not seen. This is the second time a little shearwater has been found in a Manx shearwater breeding colony, following a returning bird on Skomer, Pembrokeshire in 1981 and 1982. It has chosen a totally inaccessible site on a sheer cliff over 100 feet high and arrived in total darkness between midnight and 0100 hours.

It is known as the North Atlantic little shearwater – or the Macaronesian shearwater – and is a small shearwater which breeds in the North Atlantic.

Muscicapidae

June – Lewis, Western Isles
Collared Flycatcher (*Ficedula albicollis*)
BOU/IRBC Category: **A**
RBA Status: **Extremely Rare Vagrant**

The collared flycatcher is a rare vagrant in western Europe. It breeds in southeast Europe (with isolated populations in the islands of Gotland and Oland in the Baltic Sea, Sweden) and southwest Asia. It is migratory, wintering in sub-Saharan Africa. They are birds of deciduous woodlands, parks and gardens, and have a preference for old trees with cavities in which they nest. They build an open nest in a tree hole, or man-made nest-boxes. They are small birds at 12-13.5 cm long and take insects in flight, as well as hunting caterpillars amongst the oak foliage. It will also take berries.

Corvidae

December – Hadley, Worcestershire
Nutcracker (*Nucifraga caryocatactes*)
BOU/IRBC Category: **A**
RBA Status: **Extremely Rare Vagrant**

A nutcracker was reported briefly in a garden

at Hadley, but has not been seen since 28th December. Despite reports of this species on an almost annual basis, the last acceptable sighting was in Britain in 1998. The Spotted Nutcracker (*Nucifraga caryocatactes*), occurs in Europe and Asia. They are not migratory, but they will move out of their ranges if a pine-cone crop failure causes a food shortage.

Accipitridae

January – Sennen, Cornwall
Pallid Harrier (*Circus macrourus*)
BOU/IRBC Category: **A**
RBA Status: **Rare Vagrant**

The confirmation of the identity of a juvenile pallid harrier in Cornwall at Sennen, then later near Madron at Men-an-Tol was announced in January. This bird had been first seen in the area on 4th December but was not conclusively identified, and has proved elusive since.

The pallid harrier breeds in southern parts of eastern Europe and central Asia and mainly winters in India and southeast Asia. It is a rare vagrant to western Europe and Great Britain, but a juvenile did winter in Norfolk in 2002/03. It breeds on open plains, bogs and heathland and in the winter months it is a bird of open country.

Anatidae

January – Glasagh Bay, Co. Donegal
American Eider (*Somateria dresseri*)

The sighting of a drake American eider of the form dresseri was perhaps the first Western Palearctic record at a site which is also well known for records of borealis Northern eiders. This species of eider is usually seen on the Atlantic coast and is the only one that breeds south of Labrador. They nest from Maine northward and winter south to Long Island.

Turdidae

January, February – Glamorgan, Somerset, North Yorkshire, South Yorkshire and Buckinghamshire
April - Cleveland
Black-throated thrush *(Turdus atrogularis)*
BOU/IRBC Category: **A**
RBA Status: **Rare Vagrant**

There was news of a black-throated thrush present in Beaconsfield, Buckinghamshire in

January, making a total of five discovered in the five day period of 8th - 12th. It is thought that their presence was as a direct result of the harsh weather conditions at that time.

The black-throated thrush is a migratory Asian species and there have been 65 accepted records of it in Britain and 0 in Ireland. However, this would be the first record for Buckinghamshire.

The black-throated thrush is a common breeding migrant in East Kazakhstan; common (in some places rare) winter visitor in South and South-East Kazakhstan; and abundant (in some places rare) passage migrant in the south-east half of Kazakhstan.

Gaviidae

January – Rinville Point, Oranmore, Co. Galway, Ireland
Pacific Diver (*Gavia pacifica*)
BOU/IRBC Category: **A**
RBA Status: **Extremely Rare Vagrant**

The Pacific diver (or loon) breeds on deep lakes in the Alaskan tundra region and in northern Canada as far east as Baffin Island, as well as Russia, east of the Lena River. It may migrate in flocks and winters at sea mainly on the Pacific coast or on large lakes as far as China, Japan, North and South Korea, the USA and Mexico. It has been a vagrant to Greenland, Hong Kong and Great Britain – it was spotted near Farnham, North Yorkshire on January 2007. However this is the first sighting of this bird for Ireland.

Podicipedidae

February – Lough Gur, Co. Limerick
Pied-billed Grebe (*Podilymbus podiceps*)
BOU/IRBC Category: **A**
RBA Status: **Extremely Rare Vagrant**

by BirdLife International - the official Red List Authority for birds for IUCN): **Vulnerable**

An adult male lesser kestrel was a surprise find at Minsmere RSPB, Suffolk in March. It later relocated to nearby Westleton Heath, and was then later reported going to roost in a holm oak.

It breeds in Spain, Portugal, Gibraltar (to UK), France, Italy, Bosnia-Herzegovina, FYRO Macedonia, Albania, Greece, Turkey, Morocco, Algeria, Tunisia, Libya, Israel, Palestinian Authority Territories, Jordan, Iran, Iraq, Armenia, Azerbaijan, Georgia, Russia, Ukraine, Afghanistan, Turkmenistan, Uzbekistan, Kazakhstan, China and Mongolia. Birds winter in southern Spain, southern Turkey, Malta and across much of Africa, particularly South Africa.

The main cause of its decline has been habitat loss and degradation in its western Palearctic breeding grounds, primarily a result of agricultural intensification, but also afforestation and urbanisation. The use of pesticides may cause direct mortality, but is probably more important in reducing prey populations.

Although there have been 47 previous records accumulated between Britain and Ireland - several of which have remained for long periods - this is first record since 2003 (when one visited Tacumshin, County Wexford on 22nd-30th November), prior to which the last well-watched individuals were in 1999.

This bird breeds in south-central Canada, throughout the US, Central America, the Caribbean and temperate South America and is the most widespread of North American grebes where it is found on open waters. However, it is rare on salt water. They are year-round residents in much of their range but those that breed in areas that freeze in winter do migrate to warmer regions. It does not appear to be a strong flyer but has occurred in Europe as a rare vagrant on a number of occasions in the past and one bird in England even bred with a little grebe, producing hybrid young.

Falconidae

March – Minsmere RSPB, Suffolk and Westleton Heath, Suffolk
Lesser Kestrel (*Falco naumanni*)
BOU/IRBC Category: **A**
RBA Status: **Extremely Rare Vagrant**
2010 IUCN Red List Category (as evaluated

Apodidae

April – Folkestone, Kent, Rainham Marshes, London, and Chafford Hundred, Essex
Alpine Swift (*Apus melba*)
BOU/IRBC Category: **A**
RBA Status: **Rare Vagrant**

The Alpine swift breeds from southern Europe to the Himalayas in mountains and are strongly migratory, wintering much further south. They wander widely during migration and can often be seen in much of southern Europe and Asia. They spend most of their lives on the wing, living on insects and also drink whilst flying. They roost on vertical cliffs or walls.

Apodidae

April – Kessingland sewage works area, Suffolk
Pallid Swift (*Apus pallidus*)
BOU/IRBC Category: **A**
RBA Status: **Rare Vagrant**

The pallid swift's name comes from the Latin 'apous' meaning 'without feet' and 'pallidus' meaning pale. The 'apous' refers to this bird having very short legs and that it never voluntarily settles on the ground. It has a wide distribution over Europe and Africa, from France to Pakistan, and south to Zambia, but can migrate further south during the winter. They breed on cliffs and eaves around the Mediterranean and on the Canary Islands and Madeira and are rare north of their breeding range, although identification problems may contribute to them being under-recorded.

Hirundinidae

April – Wraysbury Gravel Pits, Berkshire
Red-rumped Swallow (*Cecropis daurica*)
BOU/IRBC Category: **A**
RBA Status: **Rare Vagrant**

This bird breeds in open hilly country of temperate southern Europe and Asia from Portugal and Spain to Japan, India and tropical Africa. The latter birds are resident but those from Europe and other Asian countries are migratory. They winter in Africa or India and are vagrants to northern Australia and Christmas Island. They construct their nests in caves or even man-made tunnels.

Glareolidae

May – Frampton Marsh, Lincolnshire
Oriental Pratincole (*Glareola maldivarum*)
BOU/IRBC Category: **A**
RBA Status: **Extremely Rare Vagrant**

This sighting was initially thought to be a collared pratincole, but the true identity was soon realised. This is only Britain's seventh record of this species - although probably actually only the fifth individual - following the bird present last spring between West Sussex and Kent on 28th May-3rd June. It is also known as the grasshopper-bird or swallow-plover and is a wader.

They are birds of open country, and are often seen near water in the evening, hawking for insects, and are found in warmer parts of south and east Asia, breeding from Northern Pakistan and the Kashmir region across into China and south west. They are migratory, wintering in both Pakistan and the Republic of India, Indonesia and Australasia. They are rare north or west of the breeding range, but, amazingly, this species has occurred as far away as Great Britain more than once. The first record for the Western Palearctic was in Suffolk, in June 1981.

On 7th February 2004, 2.5 million oriental pratincoles were recorded on Eighty Mile Beach in Australia's north-west by the Australasian Wader Studies Group. There had previously been no record of this magnitude and it is supposed that weather conditions caused much of the world's population of this species to congregate in one area.

Laniidae

May – Sennen Cove, Cornwall
Brown Shrike (*Lanius cristatus*)
BOU/IRBC Category: **A**
RBA Status: **Extremely Rare Vagrant**

This was the eighth occasion that a brown shrike was discovered in Britain and is the fourth record since 2008. The bird is mainly found in Asia on mainly open scrub habitats where it perches on the tops of thorny bushes in search of prey. Several populations of this widespread species form distinctive subspecies which breed in temperate Asia and migrate to their winter quarters in tropical Asia. They are sometimes found as vagrants in Europe and North America. It breeds in northern Asia from Mongolia to Siberia and winters in South Asia, Myanmar and the Malay Peninsula

Anatidae

February – Tacumshin, Co. Wexford, Ireland
Baikal Teal (*Anas formosa*)
2010 IUCN Red List Category (as evaluated by BirdLife International - the official Red List Authority for birds for IUCN): **Vulnerable**

Following a report of a brief sighting of a Baikal teal in January, a hybrid teal resembling a Baikal was photographed in Tacumshin in February. However a confirmed drake Baikal teal was seen a few days later. This duck is also known as the bimaculate duck or the squawk duck and breeds within the forest zone of eastern Siberia Russia and occurs on passage in Mongolia and North Korea. It winters mainly in Japan, South Korea, which held the largest wintering population during the 1990s, and mainland China, and it is a rare winter visitor to Taiwan and Hong Kong.

Molecular and behavioural studies have suggested that it has no close relatives among living ducks and should be placed in a distinct genus; it is perhaps closest to such species as the garganey and northern shoveler.

It has been classified as vulnerable due to hunting and destruction of its wintering wetland habitat. In China and South Korea, birds are killed by poisoned grain; pesticide poisoning and pollution from agricultural and household wastes are thought to be a serious problem in the Geum River, South Korea. Wintering sites in South Korea are threatened by the development of wetlands; there has been a recent proposal for building the largest tourist development in northeast Asia on the Haenem reclamation site, a key site for wintering Baikal teal·

Also, its habit of forming dense aggregations in winter renders the species susceptible to infectious diseases; 10,000 birds were recorded dead owing to avian cholera in October 2002.

Accipitridae

White-tailed (Sea) Eagle
(*Haliaeetus albicilla*)
Also known as the **Sea Eagle**, **Erne**, or **White-tailed Sea-eagle**

It was announced on 14th June 2010 that the sea eagle reintroduction project is now under threat due to Natural England reporting that it will not continue as lead partner with the RSPB in the current project for this bird to be reintroduced to Suffolk.

"The decision reflects the need for crucial savings in public funding that must be made in the current climate of financial restraint, and the need to concentrate on existing commitments to biodiversity conservation.

The sea eagle project partnership – to date jointly led by Natural England and the RSPB – has carried out an extensive programme of work to assess the feasibility of re-introducing white-tailed eagles to Suffolk. The project recently received the final piece of evidence that was commissioned in preparation for a future formal public consultation on the proposed reintroduction.

The report from the Food and Environment Research Agency looked at a variety of circumstances where sea eagles might conflict with other farming and land management interests in the potential release area, and suggested means to address any potential conflict. The report recommended a range of measures including the development of an incentive scheme that would offer payments for specific livestock management, such as extra-shepherding, planting of woodland and provision of artificial shelters. Dr Tom Tew, Chief Scientist for Natural England, said "Natural England is keenly aware of the reduced funding that will be available to public bodies over the next few years and we have taken this into account when reviewing the substantial body of evidence gathered by the feasibility study. All the evidence suggests that a reintroduction would have met all the international criteria and would have been successful. However, reintroduction of a large raptor is an expensive and difficult operation.

"We have taken the decision to withdraw from the project at this stage because we believe it would be inappropriate to commit public funds to an extensive public consultation over a project that we would, in the foreseeable future, be unable to fund. Our work to date shows both support and opposition to the project and we have listened to both sides of the debate. Whilst it will be a disappointment to many that we must withdraw at this stage, we did not think it was in the interests of the public, nor the many people involved, to delay this decision, which is inevitable given the future financial situation.

"We will honour our commitment to publish the work that the project has undertaken thus far and this may inform any reintroduction project that is planned in the future."

THE TEXAS BLUE DOG MYSTERY

Jonathan Downes

One of the greatest scenes in the whole of cryptofiction takes place at the end of *The Lost World* when Professor Challenger *et al* returned to London for a public meeting at – if my memory serves me correctly – the Royal Albert Hall. After telling of their adventures in the land of the dinosaurs, they silenced the whoops of derision which greeted their claims by opening an enormous crate from which a pteranodon emerges, squawks and flies around the room only to make an inconvenient escape through a window which some jackass had left open.

Well, before I go any further, let me tell you once and for all that – on this occasion at least – I have once again not fulfilled my role as the modern day Professor Challenger in the way that I would have liked. However, the latest CFZ excursion came back with photographs, film, scat and DNA results. But as always seems to be the way in the Fortean zoological investigations we are left with more questions than answers.

My search for the blue dogs of Texas began in November 2004 when, as chronicled in the pages of *Fortean Times*, and in my book *The Island of Paradise* (CFZ 2007) I went to El-mendorf, just south of San Antonio, at the behest of the Discovery Channel who were making the pilot of a show which was to be called *The Tracker*. When they 'phoned me up some weeks before they told me that they wanted someone to be rough, tough, no-nonsense, and all sorts of other clichés that six years later I thankfully forget. So I went there wearing a leather jacket and without having had a haircut. This was my first mistake, because I am sure in my own mind that the people who met me at the airport took one look at me and decided that I wasn't what they were looking for, and the fact that three weeks later they filmed exactly the same sequences with a bloke called Spaceman Joe who apparently wore deely-boppers on his head and spoke in a funny voice, suggests that my hypothesis is correct. However, I went to a farm where the local rancher Devin McAnally had shot an unfortunate blue skinned hairless canid in July 2004. He took photographs of it to a local convenience store owned by a charming lady called Nancy and one of the punters therein said that it looked just like "the chupacabra that her grandmother had told her about when she was a girl". Thus was born the legend of the Texas chupacabra. I took one look at the bones of the unfortunate creature and was convinced that it was nothing of the sort. However, over the previous five months the case of the Elmen-dorf beast had been discussed widely across the internet and had been roundly dismissed as a coyote with mange. Well I was pretty sure that this couldn't possibly be the case, and over the next six years I studied the matter from afar and hoped that I would eventually get back to Texas to study the problem in person.

In the spring of 2009 we became friends with a couple from Kemper, Texas who had made contact with us after reading Nick Redfern's *Three Men Seeking Monsters*. Well, as I have always said, you can believe anything that Nick Redfern or I write except what we say about each other, which should always be taken *cum gralo salis*. I admit that I started it with my portrayal of him as a leather-clad thug in *The Blackdown Mystery* over a decade ago, but ever since then we have been foisting evermore extreme depictions of the other upon our long suffering public, and – although they were too polite to say so – I am sure that Richie and Naomi West were surprised to find a happily married, and reasonably conservatively dressed bloke living in rural Devon without a monocle, deer stalker or gold lame dressing gown in sight. I hope they weren't too disappointed. I guess they weren't because we kept in touch over the next nine months and they were kind enough, not only to invite us to go and stay with them, but – with largesse which even now I find astonishing – paid for the entire trip.

It was initially supposed to be a holiday, but as both the Wests, and Corinna and I became more and more involved in the hunt for the blue dogs almost without our noticing it, it turned into a full scale CFZ investigation. And what an investigation it was!

During the planning stages Richie and Naomi went to Blanco, Texas where another specimen was languishing in the deep freeze belonging to a local student taxidermist. He took a number of tissue samples which were sent off for DNA analysis, and at the time of writing (May 2010) we are still awaiting the results.

Our first port of call was a small suburban town some miles north of Houston. Now, I hope that you will forgive me for being less than explicit with my geography here, but the lady with whom we were to visit expressly asked us to respect her anonymity. Like so many of the other cases with which we were to deal, this one had been referred both to us at the CFZ, and Richie and Naomi by my good friend Ken Gerhard, author of *Big Bird* and *Monsters of Texas* (with Nick Redfern). He had told us about the lady called Denise who lived in a suburban house with her young son and elderly mother. The house backed on to an area of wilderness owned by the local electricity company which flanked a long series of power lines and pylons which was at least 40 miles in length. This strip of wilderness was just over a mile in width and contained a rich and diverse population of animal life. Richie and Naomi had set up security cameras which picked up foxes, raccoons, deer, possums, and on one occasion a very peculiar looking canid.

Denise had been watching these strange dog-like creatures – nearly completely hairless in the summer and with a thin coat of down in the winter – for about six months and had both filmed and photographed them. One of my main arguments with the idea that the Elmendorf had been more than a coyote was that if the animal had been that riddled with sarcoptic mange that it was completely hairless, it would have been hardly able to work, let alone kill chickens, eat mulberries, wag its tail and do other doggy things. Denise's creatures were apparently able to procreate and appeared to breed true. The video footage we have of them shows them walking with a peculiar hump-backed gait, eating food off the forest floor and generally behaving like a perfectly healthy canid. Eye-witnesses to whom we spoke even had them cocking their legs and scenting trees like a perfectly normal dog.

But what were they? They had only moved into the area within the past six months and – according to Denise – the population of rabbits, opossums and other small creatures had diminished rapidly, and another wild canid – a semi-tame, but wild, coyote – appeared to be very scared of these new arrivals.

The family's own dogs, however, seemed eager to make friends and I have footage which appears to show them and the naked blue/grey dog sniffing at each through a chain link fence.

During our two days there, although we saw raccoons, foxes and opossum we didn't see any blue dogs, but we did collect scat and footprints.

We then drove west to Fayetteville where at the ranch owned by a family called Hayek we met Harvey and his son Deric (coincidentally a taxidermist). For some years they had been seeing these creatures living in several locations on their ranch. On one occasion they had found a road kill which had been sent to the local university who had been unable to identify it and who had been doing their best to buy the specimen off the Hayeks for their museum. Once again the description of the creature was surprisingly homogenous: blue or blue/grey, hairless, dog-like creatures larger than the largest coyote, with long muzzles and hunched backs. They took us to a remote part of their ranch where in the sandy walls of a desolate gulch there were a series of large holes that led deep into the sandy cliff-face. These were, or least have been, the lair of a family of these creatures, they explained. They had seen them on a number of occasions including a large specimen which went in and came out again facing the other way, which implies that inside there is an area big enough for it to have turned round.

Deric had apparently shot one six years or so before but, not realising it was of any importance, had not kept the body. The whole family had seen them and, now that they realise quite how important a discovery these animals could be, they were prepared to work with the CFZ, and with any other conservation organisations that would care to have become involved in a conservation programme.

There is a paradigm which I have noted again and again in Fortean, and zoofortean research. As people's confidence in those who are in authority over them diminishes, there is a tendency for them to blame "The Government" for those phenomena which they cannot explain. Therefore it is "The Government" who is somehow responsible for cattle mutilations, UFOs, chupacabra reports, and even in the UK some accounts of mystery cats. Similarly, as the man in the street slowly begins to realise that his species has committed unforgiveable war-crimes against the environment, environmental or quasi-environmental explanations are given for anomalous phenomena, especially those in the natural world. The Hayek family had until recent years been the proud owner of a large and fruitful orchard of pecan trees. In recent years they had seen their legacy being slowly but surely destroyed as trees withered and died and even apparently healthy trees produced little or no nuts. They blamed this upon SO_2 from a local coal-fuelled power station. Could it be these silent but deadly emissions had somehow caused an unknown mutation in one of the canids living in the area and produced these strange bald blue/grey dogs?

For it must be remembered that there are quite a few canids living in Texas. These include the red wolf (*Canis rufus = Canis lupus rufus*), coyote (*Canis latrans*) , gray fox (*Urocyon cineroeargenteus),* red fox *(Vulpes vulpes),* and swift fox (*Vulpes velox).*

The swift fox which may or may not be the same as the kit fox (*Vulpes macrotis*) is a funny little creature which is very unlikely to have contributed anything towards the genetic make-up of the Texas blue dog, but is included for the sake of completeness.

Back at Kempner we visited a church at which Naomi's father was the preacher. The music sounded like it came from a Flying Burrito Brothers songbook and one wished that the building had been a tar-papered shack as described by Truman Capote rather than a modern, purpose-built community centre, but one can't always get what one wants. After the service we met a young man called Jesse who had seen one of these strange bald canids a few years earlier at San Antonio airport. He described something which was becoming excitedly familiar: a hairless, blue/grey dog-like animal that was scampering about minding its own business, and not giving any indication that it was suffering from a debilitating and potentially fateful disease. Interestingly, however, it didn't seem to take any notice of the noise despite being on the campus of what I know from personal experience is a particularly noisy 21st Century airport.

Our next port of call was to my old acquaintance Devin McAnally. Devin is a very complicated man and has been royally screwed about by nearly everyone with whom he has had to deal. I am not one of those people, but I think he has been too badly scarred by his contact with the multi-headed hydra which is the modern multi-media corporate monster/machine, and – sadly – I am afraid that he feels that I and the CFZ are amongst the people who have treated him badly. This is actu-

ally not true. I have nothing but respect for the man: he has been planning a resource for under-privileged children for many years, is a highly intelligent and gifted academic, and is one of the few men who – during my years as a crypto-academic – I feel privileged to have met.

Our visit there added little to our knowledge of the Texas blue dogs. We were to return 36 hours later, and the two visits have concatenated themselves together in my mind. However, whilst in Elmendorf we learned a lot both about the local cryptozoology and about the nature of the Fortean universe itself. We heard that the Elmendorf dog shot by Devin in 2004 was not unique – other specimens have been seen, and even shot, but there were no other bodies available. However, there were several other mystery animals that had been reported from the area, including an unknown species of wildcat which was described by Devin as a bobcat with a long tail, and sightings of "panthers" which were presumably surviving eastern pumas. Until about 2000 it was widely believed that there had been three species of the puma (aka mountain lion; cougar) in North America. These had been the western puma (*P. concolor puma*) , the eastern puma (*P.c. linnaeus*), and the Florida panther (*P. c. coryi),* the last of which is a massively rare creature numbering double figures only, reported from the Everglades. However, in about 2000 they were all lumped together as the North American puma (*Puma concolor couguar*). This decision may or may not have political implications: both the eastern puma and the Florida panther were counted as critically endangered with the former being presumed extinct across nearly all of its range. It had indeed become a cryptid, with several books (the most recent written by Jay W. Tischendof and Bob Butz) written about it. It may be cynical of us here at the CFZ but it cannot be denied that one of the side-effects of taxonomic "lumping" is to remove legislative protection from various creatures. Whilst the eastern puma was seen as a separate subspecies, should it have been found and saved from extinction, then it would have been an expensive matter indeed to have afforded it federal protection. Now it is lumped together as one subspecies, both regional and national governments will have far less to worry about. However, whatever its taxonomic status, pumas were allegedly extirpated in Texas a long time ago, and the fact that Devin McAnally and his friends are convinced that they still exist is of major zoological importance.

One of my favourite authors, the late Sir John Verney, once wrote that if you spent enough time sitting outside the Uffizi Gallery in Florence you will see everybody you know walk passed you. I am not sure that that is true, but one of the biggest truisms of a Fortean universe is that one meets friends and acquaintances in the oddest of places. Waiting to meet us at Devin's mobile home in Elmendorf were two young men who would not have looked particularly out of place in an episode of *The Dukes of Hazzard*. Chad and Jonathan were two of the nicest guys that we had met in a long time and knew one heck of a lot about the local wildlife, telling us many things which helped fill in our knowledge of the local fauna. One of my oldest friends is a geezer from Bristol, who as well as being a reasonably well-known record producer has also been involved in the fringes of the music business as a purveyor of class B listed substances. To find that 'Jonathan' had once been tour manager for the late Paul Raven (ex Killing Joke) and was a good friend of my quondam dope dealer underlined for us (if any underlining was necessary) that the omniverse is

a ridiculously complex place, and that the cosmic joker has an infinitely subtle sense of humour.

And then it was to Cuero.

Dr Phyllis Canion was obviously the lady of the manor – she reminded of my late mother. She traversed through the genteel little country club like a battleship in full sail and with the noblesse oblige of Penelope Keith in *To the Manor Born* (with a little touch of Princess Margaret meeting the Rolling Stones) it had been arranged for us to meet Dr Canion as a result of her appearance on a National Geographic documentary about the Texas blue dogs. Dr Canion (whom – with a nod to Georgette Heyer's *Grand Sophy* I had dubbed 'the Grand Canion' in my own personal rollerdex) was the most social impressive of the witnesses/pundits that we were to meet. Some years earlier she had (in her obvious persona as lady of the manor) come across four individual Texas blue dogs; all male, all identical, and all road kills. Through misadventures two of them had fallen by the wayside, but the remaining two carcasses had been preserved by her.

It was, I believe, Oscar Wilde who said that Great Britain and the United States were two great nations divided by a common language. The National Geographic documentary had likened the Cuero blue dogs to the late lamented *Thylacinus cynocephalus* and had even described how – alongside a single pair of nipples – were "pouches" on the back legs. Corinna and I, as Brits, had immediately interpreted this as insinuating that the mysterious blue dogs were marsupials. Over an excellent dinner, I tried to draw out Dr Canion on the subject. Would she discuss the matter? Would she heck!

No matter how many conversational gambits I tried, she seemed determined to ignore the topic of Texas blue dogs, and instead talked (eloquently and entertainingly) of everything and anything else. Although I was getting massively frustrated, and more than a little cross, I did have probably the best meal I had ever eaten in the New World in front of me (I highly recommend liver and onions cooked with mesquite and sweet potato chips, with a green salad) and I ate my food and drank my wine politely, and with great gusto.

When we finally made it to Dr Canion's ranch-style house all was revealed. For there in her fireplace was a stuffed and mounted Texas blue dog. She burst out laughing: "I wanted y'all to see this for yourselves, and to see your faces," she said.

But was it a marsupial? Could it, as some internet pundits had suggested, be a peculiar example of convergent evolution? A new world thylacine analogue having evolved from the carnivorous opossums of North and South America?

No, of course it wasn't.

The first thing I did was to have a look at the much vaunted "pouches", and this is where the blessed Oscar's linguistic advice came in ever so handy. Every English person that I know that has seen the documentary, or read comments about it on the internet, had assumed that the "pouches" referred to blithely were those trademarks of the marsupials; protective membranes under which the semi-developed ur-foetus which is ejected unceremoniously from its mother's birth canal long before it is able to face the rigours of the outside world, can finally develop. And of course, they were nothing of the sort.

When Dr Canion and others referred to "pouches" they were referring to things that looked like bulging packets of meat, roughly the shape and size of a large raisin scone which were plonked on the haunches of the animal, roughly where its buttocks would be, if indeed it had buttocks (which it doesn't). My immediate thoughts were that these were anal glands, or some rough analogue of same. However, they weren't. I questioned Dr Canion in some depth about these, and she insisted - much to my surprise - that they were flesh, and not glands of any sort. When one took a closer look at them, it was obvious that there were not anywhere where there should be any glands of any kind on any self-respecting canid.

It was a remarkable creature, and apart from the "pouches", and for the moment we will continue to call them "pouches" because we cannot think of any better name, there were four other notable features.

It was almost completely hairless, and whilst there were hair follicles on the skin they were few and far between. Dr Canion insisted, and we see absolutely no reason to disbelieve her, that she investigated the hair follicles of the recently dead creature and found them to be perfectly healthy.

Like Hitler, and my late dog Biggles, the specimen was apparently monorchid.

The eyes were a remarkable pale blue. I would have taken exception to this, and assumed that it was the result of incompetent taxidermy, but Dr Canion showed me a photograph which proved that this colour was exactly the same as the eye colour in the recently dead animal.

It was mounted in a peculiar hunch-back

position. I queried this with Dr Canion, and she confirmed to me that when she has seen the specimens of these animals alive, they have stood in this very manner.

We recorded several hours worth of interview with Dr Canion, and that night as we drove back to our motel in the knowledge that the investigative part of our trip was largely over, we pondered deeply over the remarkable events, not only of that day, but of the previous two and a half weeks.

So what the bloody hell are they? That is, of course, the ultimate question. Bernard Heuvelmans himself introduced cryptozoology to a remarkable concept within cryptozoology, that of a cryptid with multiple identities. He cited the chemosit Nandi bear; a semi-mythical and very fearsome predator once recorded from various parts of East Africa, and even now occasionally reported. He suggested that there were several different animals responsible for sightings and encounters with this "creature", and that the Nandi bear was in fact a composite of encounters with *known* animals such as rattels (honey badgers), aardvarks, and out of place chimpanzees, together with sightings of *unknown* animals such as late surviving chalicotheres and a hypothetical species of giant baboon. Since then this multiple identity syndrome has been used by many cryptozoologists including myself to explain several of the world's most well known cryptids including the Loch Ness monster, and so I make no apologies for – to a certain extent at least – resorting to it here.

It is certain, and unusually within cryptozoology I *can* say certain, that not all of these blue dogs are of the same species. Genetic material from the Elmendorf creature was tested at two laboratories: one in New York and one, thanks to Lars Thomas of the CFZ, in Copenhagen. Both tests proved conclusively that this animal was a domestic dog (*Canis lupus familiaris*). However, five different tests on the Cuero creature all identified it as a cross between a coyote (*Canis latrans*) and a Mexican wolf (*Canis lupus baileyi*).

Herein lies the problem. In fact herein lie several problems.

Although the Mexican wolf (*Canis lupus baileyi*) was once found in Texas, as far as we are aware its range *never* included Cuero, or indeed the other areas that we have been investigating. It was Richie West who noted that the blue dogs have appeared along what he has dubbed the 'highway 10 corridor' following one of the main roads of Texas. But although *Canis lupus baileyi* was never – as far as we know – found in this part of the Lone Star State, the Texas grey wolf with the monumentally fortean Latin name of *Canis lupus monstrabilis* was once found across this part of the state. Yes, gentle reader, you have noted my past tense for according to accepted wisdom the last Texas grey wolf was shot in 1942. Another sub-species, the buffalo wolf (*Canis lupus mubilis*) once followed the bison herds across the state's plains including central and southern Texas, although the last of these was shot in 1926. However, this is where it gets complicated.

A few years ago wolf taxonomy was revised and twelve of the original sub-species which occurred in the western United States and central Canada were reclassified as *canis lupus mubilis,* so according to some taxonomists the buffalo wolf still exists although everyone agrees that it no longer exists in Texas.

The status of the Mexican wolf (*Canis lupus baileyi*) is also on shaky ground. The last two Texas specimens were both shot in 1970, and in a rare display of co-operation between the American and Mexican governments, the last five wild Mexican wolves were allegedly captured in 1980, and used to start a breeding project. Several hundred have been bred in captivity, although from an extremely limited gene pool, and a hundred were liberated in southern Arizona. However, by the time we were in southern Texas only forty-two of them were left, and they were over a thousand miles away from Cuero. It seems highly unlikely that a wandering male from this population could have sired the Cuero creatures.

There are suggestions that a relic population of *baileyi* still exists in the Sierra Madre and during our sojourn in Texas we discovered a surprisingly large number of anecdotal accounts of wild wolves being found in several locations in the state. At the very least this would suggest that either a small pocket of *baileyi* still exists in the wild, or that *monstrabilis* managed to evade extinction. Even if these animals turn out to be surviving *nubilis* then the existence of living genetic material from the buffalo wolf could well cause the taxonomic revisions of a few years ago to be looked at again. But it gets even more confusing. Because – depending on who you believe – there is a second species of wolf in Texas. The red wolf (*Canis rufus*) was supposed to be extinct in the wild, but our friend and colleague Chester Moore Jnr. rediscovered them in the late 1990s by using camera traps set in his native Orange County. But is the red wolf a separate species? Well, once again, it depends on who you believe.

It is generally believed that the modern canids (which include dogs, wolves and jackals, but not foxes or maned wolves) evolved in southern Texas (of all places) something like 50 million years ago and that the red wolf (*Canis rufus*) evolved in Texas about 1 million years ago spreading northwards and into Eurasia across the land bridge where – probably in the near East – they evolved into the ancestors of the grey wolf which returned to North America by the same route between 600,000 and 300,000 years ago. Therein lies another condundrum; is the red wolf (*Canis rufus*) generically distinct enough from the grey wolf (its descendant) to be considered a separate species or should it be considered merely a subspecies albeit a primitive one of the grey wolf, and be correctly named *Canis lupus rufus?* Those of us who believe that it is a separate species, have it, the coyote, and possibly the eastern wolf (*Canis lycaeon*) (an animal whose specific identity is even more controversial than *Canis rufus*) as *bona fide* North American canids, whereas – of course – *Canis lupus* originated in the Old World. Confused? So would anybody be. I have been living with the phylogenetic tree of the North American canids all year, and believe me it doesn't get any easier.

It was Naomi who first noticed that several of the photographs of dead blue dogs from across southern Texas that had been collected by our friend and colleague Ken Gerhard show the creatures exhibiting the "pouches" that are such a singular feature of the mounted Cuero specimen. Indeed, when you look hard enough, even some of the animals filmed and photographed by Denise also feature these peculiar features on their nether regions. However others do not. All the animals that feature the "pouches" appear to be male. Could this be an example of sexual dimorphism? Or are the animals without "pouches" something else entirely?

The Elmendorf beast from 2004 had no "pouches". But it was a female. The DNA tests reveal it as being a domestic dog, and without access to a complex reference library of genetic material it is apparently very difficult and expensive to go any further in investigating what domesticated race could have been the progenetive of this unfortunate creature. At the time of Columbus it appears that there were a large number of native American hairless dog breeds, a small number of which have survived to the present day. Is it possible that one of the supposedly extinct breeds has resurfaced due to its genetic legacy surviving unsuspected in the feral dog population of the Elmendorf region? Yes, quite possibly. However, even with the Elmendorf creature, the truth is (once again quoting Oscar W) neither pure nor simple. Remains of two different animals were jumbled together in the same plastic bin bag by a TV company who shall remain nameless, and so it is not possible to be 100% sure which skull and bones are that of the Elmendorf creature and which are of some feral dog or other. Only months after I first visited Devin McAnally I was sent a photograph by somebody who really should have known better who insinuated that this was of the Elmendorf creature. I published it as such in good faith both in *Fortean Times* and *Animals & Men.* Six years later it turns out to be nothing of the sort, but a photograph of an unknown canid from Arizona which may or may not have been suffering from an unknown skin condition.

Ken Gerhard and Naomi West both together and separately have done a remarkable job in collecting together several dozen photographs of blue dogs, mostly dead. I agree with Ken Gerhard in that a large proportion of these, and several of the so called Texas chupacabra videos on the internet, are of nothing more than very ill mangy dogs or coyotes. However, as you have seen, a small proportion including those secured by Dr

Canion and filmed by Denise (and by Richie West at Denise's property) are – I believe – something of far more importance.

From the available evidence at the very least they show that wolves are not entirely extinct in Texas, and it is - we hypothesise – not at all beyond the bounds of possibility that the discovery of these wolves may have enormous implications for the survival of the rarest sub-species. The Elmendorf creature is something else entirely and Devin McAnally should be congratulated for his patience in putting up with being jerked around by so many TV cameras. Whether or not it is a surviving member of the pre-Columbian domestic races of dog we may never know.

We are still awaiting the results of the DNA tests on the genetic material taken from the Blanco beast but would make a educated guess that it will prove to be the same as Dr Canion's specimen, and that the morphological peculiarities of both beasts are similar enough to suggest that the differences are purely sexually dimorphic. We are awaiting these results, and any to come from the Fayetteville creatures now or at any time in the future with interest. Here we should probably make brief mention of the creature filmed by a police car in DeWitt County, a figurative stone's throw from Cuero. This did not appear to have the buttock "pouches", but had a peculiarly elongate muzzle, and appeared to have the diagnostic hunched back of the Cuero and Blanco beasts (as did Denise's animals). We would hazard a guess that the DeWitt creature was probably a female, as it was far too energetic and exuberant to be merely a diseased mutt or coyote.

But what of the alleged vampirism? I have hunted the chupacabras in Puerto Rico on two occasions and have published my conclusions explaining away its alleged vampirism. My 1998 sojourn in Central Mexico produced evidence for vampiric attacks on livestock that was far less easy to explain away, and although I refer anyone interested to my 2001 book *Only Fools and Goatsuckers*, I would suggest that the events of the summer of 1996 which I have described in the aforementioned book have absolutely nothing to do with vampiric predations allegedly performed by the blue dogs of Texas.

There are three incidents which have (as far as we are aware) been the only ones to lead to the allegations that the Texas blue dogs are somehow linked with the chupacabras of the Canovenas plateau of Puerto Rico.

1. The unnamed old Spanish lady in Nancy Delio's Corner Store in Elmendorf, TX who allegedly proclaimed that this was the same thing as referred to by her grandmother.
2. The large number of poultry killed by the Elmendorf beast which prompted Devin McAnally to kill it in July 2004. He told me that so many were dead and apparently uninjured that he could only suppose that they had been killed for their vital fluids. The history of British fox hunting is full of accounts of how the British wild canids kill indiscriminately but only carry off a tiny percentage of their victims for food. It seems not impossible that what is *de rigeur* in the British farmyard can also happen in the wilds of Elmendorf.
3. Dr Canion told us of an account of poultry that *had* been exanguinated, and furthermore the incident occurred at the same time as *her* blue dogs were in the area. I have been studying the behaviour of various mustelids for many years, and whilst I

would not wish to be impertinent and contradict Dr Canion, the attack she described to me sounded very much like the ones I have encountered in the UK when domestic poultry had been predated upon by weasels (*Mustela nivalis*), stoats (*M. erminea*) and polecats (*M. putorius*). In Texas there are 11 known mustelid species, at least one of which – the North American mink (*Mustela vison*) is well-known for predating domestic livestock by drinking their blood. I am deliberately not using the word 'vampiric' because of the pejorative nature of the term. At least two other species – the black-footed ferret (*Mustela nigripes)* and the long-tailed weasel (*Mustela frenata*) would – in my humble opinion – be likely to feed upon domestic livestock in such a manner.

I do not know enough about the biology of the six species of skunk to be able to comment on whether they are likely to feed in this manner, but I have vague memories of such things, although I have to admit I have not been able to find citations for the purposes of this article. Finally the ubiquitous stoat is found in Eurasia and North America north of the 40th parallel, and if the aetiology of the species is the same as in New Zealand where it is an invasive predator introduced for some ridiculous reason by colonising Brits in the 19th Century, then it might well be present in the Cuero region without anybody realising it. Those people who are interested in further exploring the blood sucking activities of British mustelids are referred to in my 1996 book *The Smaller Mystery Carnivores of the West Country* (CFZ Press).

In 2004 I made my second visit to the island of Peurto Rico where I rekindled my acquaintance with Ismael Agayo of the Canovenas civil defence. Nick Redfern,

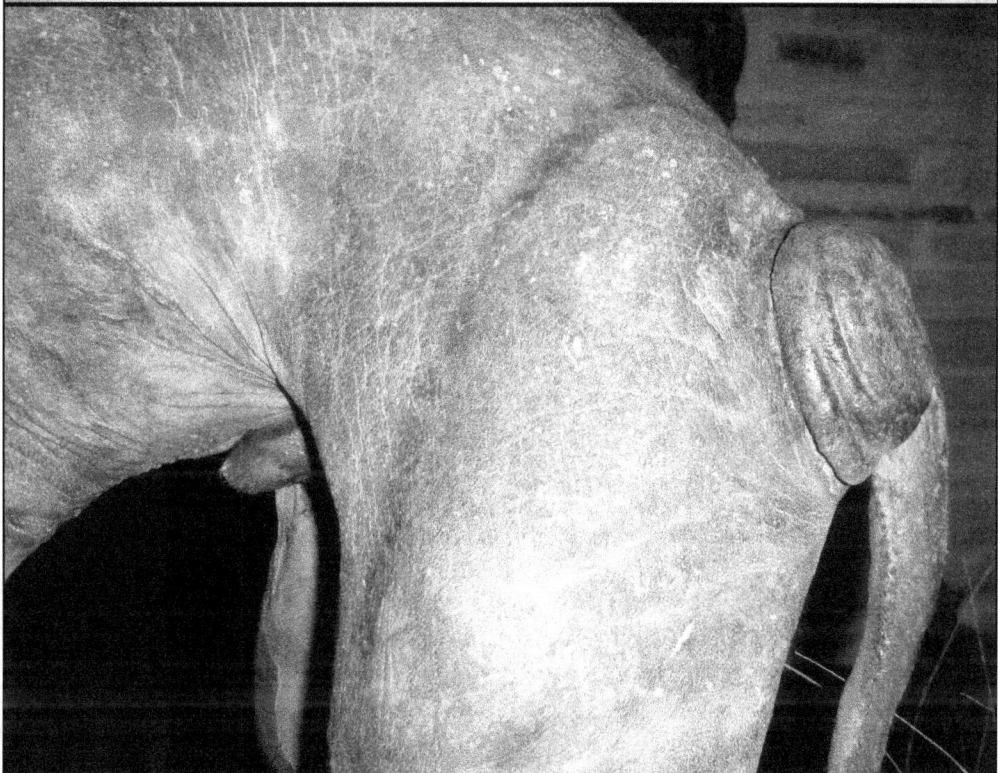

Two frames from the footage taken at Denise's house. The bottom one has an inset to show how the 'pads` on the haunches of the animals are present in the living creatures, and not some kind of taxidermic construct.

Ismael and I sat in a bar sheltering from a torrential rain storm and talking about the history of the term "chupacabras". Ismael told me that he had coined the term back in 1994 when investigating the first vampiric attack on pygmy goats in the region.

My Spanish is appalling. I can just about ask for a beer (*dos cerveza por favor*) and buy a packet of cigarettes (pointless these days as I have given up), but cannot discuss the minutae of semantics. Despite what I have been saying in lectures, articles and in my book *Island of Paradise* for the last six years, I think that what Ismael actually told me was that he and his pal had decided to apply the name of a pre-existing entity from Hispanic folklore to the spiky backed predator of the Canovenas plain. I think that the unnamed old lady in Delio's Corner Store did exactly the same thing. The phenomena in Puerto Rico and the "Highway 10 corridor" of southern Texas have nothing more in common that names given to them for totally different reasons. Certain sectors of the cyrptozoological community jumped to one conclusion, and I jumped to a totally different conclusion. We were all wrong.

I have said for years that the scientific community should stop being so as arrogant to presume that they have all the answers as to the make up and function of the omniverse at the beginning of the 21st Century. These latest revelations as to the origins of the "chupacabras" proves that the cryptozoological community, and Jonathan Downes in particular, should not fall into the same trap. As to the specific identity of the southern Texas blue dogs, watch this space. The game, is still very much afoot!

CRYPTIC SUPER-MOLES

Max Blake

In nature there exists a strange niche within ecosystems: the small burrowing carnivore. In Europe, this is represented by the familiar European mole, *Talpa europaea*, but other lands have far more obscure species, both from modern times, and back in deep geological time. In North America, a wider range of moles exist than in Europe, from the odd shrew-mole, *Neurotrichus gibbsii*, to the very odd star-nosed mole, *Condylura cristata*. Australia has its totally unrelated marsupial moles, *Notoryctes*, whilst Africa has both golden moles, and mole rats (again, both groups are unrelated to the others (also, mole rats are not carnivores, but show a strong convergence otherwise)). South America has its pink fairy armadillo, *Chlamyphorus truncates*, and used to have the marsupial *Necrolestes*. But, the oddness of all of these species is dwarfed by what I would put forward as a candidate to the title of the strangest mammal ever, *Proterix*.

Though known from a number of skulls and vertebral sections from the late Oligocene to the early Miocene, limb bones, shoulders or a pelvis have never been found. The spine is strong, but can only undulate in the vertical plane, which together with thick armour plating on the front of the skull would have provided a powerful digging device for the critter. The implication that it had no limbs is incredible, but it could be that by total fluke all fossils of the animal lost their limbs, or that they were so small when alive that they broke apart easily compared to the skull and vertebral column. But still, the strong implication that *Proterix* is the only mammal known to have totally lost all of its limbs stands it out as an incredible animal.

Which brings me nicely to some other mole-like organisms. African golden moles are part of the clade Afrotheria, putting them alongside elephants, tenrecs, sirenians, hyraxes, sengis and aardvarks in the evolutionary tree. Uniquely, they have been described as the world's only terrestrial swimmers; they all live in sandy deserts and due to the flowing nature of sand the "moles" cannot create burrows because they fall down behind the advancing mole, thus they swim through the flowing sand. Watch a video of one and you can see this process clearly. They make long trails behind them which renders their presence generally obvious and the trails are far more commonly encountered than the

moles themselves.

We have seen that one of the world's strangest mammals is known from multiple specimens, so what are we to make of Van Zyl's golden mole? *Cryptochloris zyli* was known for years from a single specimen from north western South Africa and was described in 1938 by Shortridge & Carter. This was all that was known of *C. zyli* until 2003 when another specimen was collected about 150km further north along the coast. That is the extent of the scientific evidence for this species existence. No-one has ever seen the animal alive and reported it to a scientific establishment. Inferring from what is known about the other *Cryptochloris* species, Van Zyl's is around 80-90mm long and weighs between 20-30g when adult. It is likely to be a ferocious carnivore for its size, eating invertebrates, small mammals and lizards.

The IUCN lists it as Endangered (3.1) with tourism and mining affecting its habitat, but notes that there is so little information on it that no one really knows how common it is. To all intents and purposes, it is an incredibly obscure little animal from an incredibly obscure group.

Chrysochloris visagiei is another golden mole, this time known from only one specimen from northern South Africa. It was named in 1950 and although a number of expeditions have set out to the area, none have been successful in capturing another specimen. Agriculture in the area has changed the landscape dramatically, and it may be that this species is already extinct.

Moving briefly to true moles, the Senkaku mole (*Mogera uchidai*) is an obscure mole from the Senkaku Islands north east of Taiwan. The islands are uninhabited with a mere 2.7 square miles of land area collectively. It was named in 1991 by Abe, Shiraishi & Arai from a single specimen collected in 1979 from the largest island, Uotsuri-jima. A survey from 1991 failed to find any moles, but did find a tunnel. Introduced goats threaten its survival, as does the conflict between Japan, Taiwan and China over territorial claims.

But wait, could it be that there is an even more mysterious "mole"? Well, yes. The Somali golden mole (*Calcochloris tytonis*) is my candidate for the least known, known (in that it possesses a fully valid scientific name)

animal. It was found in 1964 as a partial skeleton in an owl pellet in Somalia. That's it. No sightings, no photographs, no skins, no feeding traces, virtually nothing exists for this species, apart from a fragmented skeleton.

Regardless, this skeleton is rock solid proof of its existence, despite being backed up by nothing at all. The IUCN listed it as Critically Endangered for a while, then downgraded it to Data Deficient due to the fact that little sampling has taken place in the area in which the pellet was found, which is a good enough reason to move this enigmatic species to Data Deficient, so long as *someone* actually goes out and does some sampling of this area. Should they find it, they would be the first people ever to see this wonderful little animal.

OK, so perhaps the title of this piece is a little far of the mark: none of these animals are cryptids in any sense.

Like the Texas blue dogs and non-native British cats, both being categories known from multiple specimens, skins, sightings, photographs and video footage, these animals are not cryptids because they are not just known from anecdotal evidence. So why do many cryptozoologists know about and study blue dogs and exotic cats, but have no idea about these incredibly enigmatic golden moles? It's because they are not monsters.

No-one is scared by a ball of fuzz three inches long. The idea of a large cat equipped to kill you with a single bite living in your local woods is far scarier than a small cuddly insectivore (not in the taxonomic sense) which spends most of its life zooming around under the sand or earth.

Letters to the Editor

The Editor and his band of merry men (and women) welcome an exchange of correspondence on any subject of interest to readers of this magazine.

We reserve the right to edit letters and would like to stress that opinions voiced are those of the individual correspondent rather than being necessarily those of the editorial team or the Centre for Fortean Zoology. Every attempt is made not to infringe anyone's moral rights or copyright, and we apologise if we have unwittingly done so.

Modern Blues

Dear Editor,

I have just received my copy of *Animals & Men* – issue 47, with some relief: it has been such a long time since I have heard from the CFZ that I had become concerned that my attempt to pay membership by direct debit had gone badly awry!

I was particularly interested in the Expedition Report – Four Go Mad in Sumatra: p47-57. I wonder if I may offer some (amateur) observations?

The print produced on p51 in the sketch pad, looks very similar to that produced by a gibbon, with a long narrow heel and a very divergent big toe (see Shultz, 1969 p89). In addition the sketch of the Orang Pendek with its narrow waist and broad shoulders is not inconsistent with a Siamang Gibbon. These animals are mentioned as being present at the site. (The description of the line of darker hair down the back and the way the animal is depicted "hugging" the tree is, however).

I am surprised by the sheer quantity of Orang Pendak tracks observed by the team on Gunung Tuju. I am also intrigued by the description at the area where the tracks were observed, as being "more open with larger trees".

The writer does not mention wether the canopy was closed or the distance between the individual trees. I wonder if the Orang Pendek – at this site at least – might be a population (or possibly even an individual), of Siamang gibbons that for some reason or other, have become more terrestrial?

Gibbons are famous for their ability to walk bipedally. In zoos, at least, they seem as happy on the ground as on climbing equipment. In addition, Siamang gibbons can appear surprisingly large – almost chimp like. Perhaps the "treescape" at this site mitigates against extensive brachiation and the animals are having to use the ground to access feeding or sleeping sites. To do so, they would walk bipedally.

I have been unable to locate any quantative data about the extent to which gibbons may use the ground, other than the general "… they descend to the ground rarely." The discovery of a population of Siamangs that were using the ground to a greater degree that previously thought, whilst not being so dramatic as an unknown species of Ape, would still be very important.

It is not beyond the realms of possibility that where the observed tracks do differ from Gibbon tracks, that difference might be the result of some evolved morphological changes resulting from this increased use of bipedalism.

I believe the feet of highly terrestrial mountain gorillas are different from the more arboreal lowland forms. It may also shed light on the theory that bipedalism in humans, arose as a result of adaptations to vertical clambering and brachiation.

Regards,

Martin Knight
Bedford
11th February 2010

Ophelia

Dear Jon and Corinna,

I wonder if you can help us. What on earth is this? We found it in the corner of a field in North Cornwall last summer. We had pulled into a layby to answer a call of nature, climbed over a field gate, and there it was. Bizarrely, next to it was a pair of Wellington boots.

What do you think this is? An unknown animal, a bizarre art project, or a dead sheep?

Recently we went back to see if we could recover the skeleton for you, but it had gone.

It is a pity because we thought that it would have made a great exhibit in the CFZ museum, and then perhaps someone at the Weird Weekend could have identified it.

All the best,

Dougie and Jools,
Cornwall

The keen-eyed amongst you will have noticed that the Letters Page headlines don't make much sense. This is because they are always songs from whatever album we are listening to at the time. Last month's record was Embryonic (the gloriously experimental new record from *The Flaming Lips*..)

Can you guess what record we have pillaged in *this* issue?

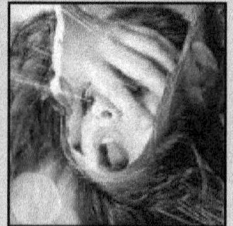

The skeleton found by Jools and Dougie

REVIEWS

David Waldron & Christopher Reeve, *Shock! The Black Dog of Bungay,* London, Hidden, 2010

142 pp illustrated, photographs, bibliography ISBN 978 0 9555237 7 9
Approx $15 US and £8.50 UK

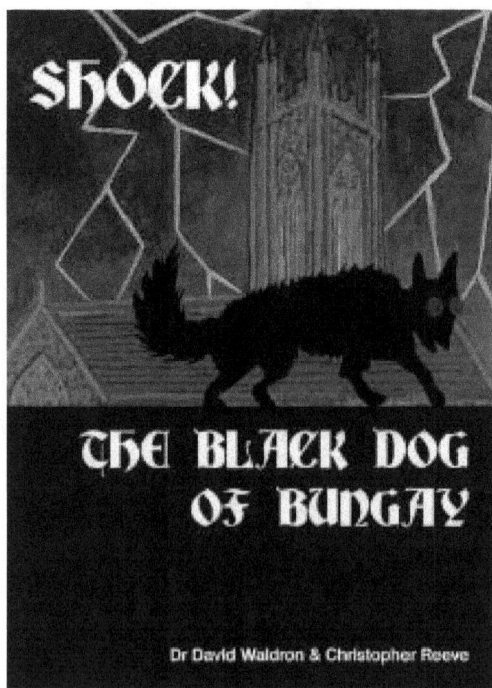

In 1577 in Bungay, Suffolk a spectral black dog burst into the town's church during a massive and unseasonal thunderstorm and savagely killed several people in ways that were unexplained, and (by contemporary descriptions of the event) positively demonic. In nearby Norfolk there is similar legendary beast, Black Shuck.

In the aftermath of the Reformation, social and religious upheaval was prominent and the dog can be seen as a symptomatic motif, a metaphor for the challenging times. Or is that all? Dogs are in many respects simply wolves who have relaxed a little, and they have the power to fascinate and terrify still, and especially *black* dogs seem to be heavily associated in folklore with storms and the Devil's work. The authors explore the subject with vigour, exposing previously unknown facets of the tale (including recent 'sightings' of portentious black dogs) and producing a rounded and wholesome book, whose depth, breadth and detail of research is belied by the page count- many books twice the size do not have half the content.

A fine example of both the internationalisation of collaborative research potentials (David Waldron is an academic in Australia and Christopher Reeve is a local historian in Bungay). This in an exemplary modern history book that is both micro- in terms of a location and a particular story, and macro- in the way it deals with the very process of historicising, and the way that we all make meaning from our collective pasts, and use legend, myth, story, motifs and symbols in our everyday lives- the black dog appears in numerous local business logos and has even made it into a reader directly and intelligently but without use of unnecessary jargon and is a credit to the field of study. The authors are to be congratulated on producing an intelligent and enchanting history book that will be of broad appeal to historians, folklorists, anthropologists, cryptozoologists, occultists and lay readers alike.

Thoroughly recommended. **Dr Dave Evans**

Taylor, Greg (editor), *Darklore Volume IV,*
Daily Grail Publishing

This forth volume of Greg's highly respected journal of esoteric science read like a who' s

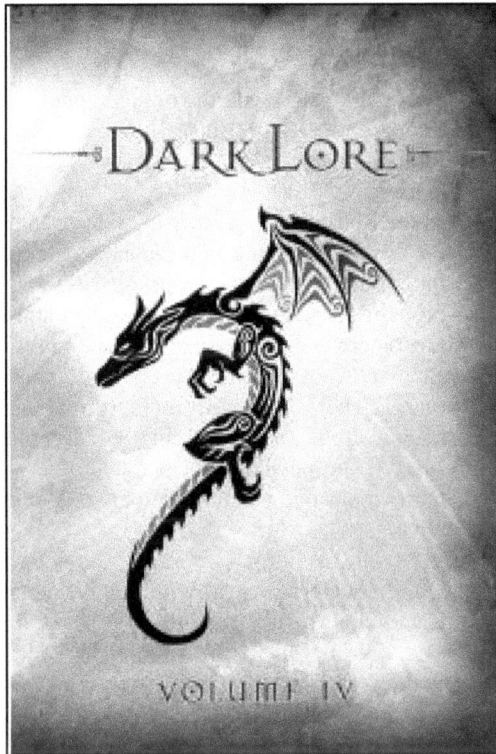

who of contemporary forteana with papers from many of today's leading researchers into 'damned' subjects.

It is a mixed bag of subjects that will appeal to both fortean generalists and to specialists in particular fields as well as the curious layman. There are fifteen papers covering most areas of forteana.

Pick up most general books on fortean subjects and you will see the same old subjects re-hashed time and time again. Not so with *Darklore*. It consists of seldom written about subjects or familiar subjects looked at in a new light or angle.

Here in we have such delights as Neil Arnold's look at blood sucking beasts prior to the chupacabra flap and finding truly weird looking beasts lapping up vital juices all across the globe (and not one of them looked like a cast member of 'Twilight'). Neil's writing is always a pleasure to read as he turns up such obscure material unseen by the beady eyes of forteans. Theo Paijmans has a similar knack of finding accounts that seem to have bypassed most other researchers. In his article 'The Newhallville Terror' he follows a spate of sightings of what appeared to be none other than Spring Heeled Jack having immigrated to the US and terrorizing a small Connecticut town.

Nick Redfern delves into government experiments into animal ESP predictably for espionage. Many experiments were apparently carried out on both sides of the iron curtain including a cyborg cat developed by the CIA for spying on the Kremlin. It cost the American taxpayers over $10,000,000 and was run over and killed by a taxi on its first mission!

John Reppion investigates the Liverpudlian eccentric Joseph Williamson known to locals as The Mad Mole'. Williamson spent most of his adult life constructing a huge subterranean labyrinth beneath Edge Hill. His story recalls that of the Winchester House wherein Sarah Winchester spent her late rifle manufacturing husband's fortune on building a never ending house. Williamson employed half the able bodied men in the area to work digging his tunnels for decades. Most think it was just philanthropy towards the unemployed but John Reppion thinks he may have had another motive.

Other treats include a refreshingly sceptical look at carvings of 'helicopters' and 'flying saucers' in ancient Egypt, a look at the world's oldest temple in Turkey, and my own humble addition, an overview of modern reports of *yokai* (monsters) in Japan, a pre-curser to my book on the subject. *Darklore IV* will appeal to even the most jaded of forteans but is also a stimulating gateway to the novice. **RICHARD FREEMAN**

Forth, Gregory, *Images of the Wildman in Southeast Asia—an anthropological perspective*
Routledge

Every so often a cryptozoological book comes along that is destined to become a classic of the field. It's a rare event but worth waiting for. *Images of the Wildman in Southeast Asia* is a case in point. The book concentrates mainly on Indonesia and is unquestionably the most detailed and comprehensive work of its kind ever writer. Gregory Forth has really put an impressive amount of work into this book not only in visiting most of the areas in question repeatedly for over twenty years but in delving into old records, foreign journals and books and even unpublished papers and works.

The main subject is the Ebu-gogo of Flores and the detailed account of their destruction at the hands of local tribes men who were supposed to have trapped them with burning plant material in a cave. This same legend is found in other areas of the island and also in other parts of Indonesia and beyond. Forth examines these cases and looks at possible links with real events like volcanic eruptions. The destruction of the Ebu-gogo, if based on a real event may have happened less than 200 years ago! He also looks at alleged modern day sightings and the possibility that Ebu-gogo may have been one in the same as the tiny hominid *Homo floresiensis*.

There is a chapter on the orang-pendek of Sumatra, a creature that I have searched for several times. Forth concludes that it is probably a real animal but not the same as the Ebu-gogo. I heartily agree with him on this point as the orang-pendek seems much larger and more primitive. It is probably a great ape rather than a hominid. There is much unseen material in this chapter. A lot of it is from Dutch colonial times. There are also unpublished accounts from the researches of Debbie Martyr, the person who, perhaps, knows more about this creature than anyone else alive.

Other chapters look at wildmen on other island chains, on mainland Asia, Australasia, Europe and North America. Sadly South America, a continent rich in Wildman lore is excluded for reasons of space.

None of these other chapters go into as much detail as the ones on Indonesian wildmen.

One of the few marks against the book is Forth's general dismissal of larger wildmen. This seems very illogical. He tries to write off the yeti for example as best explained by a bear. Anyone familiar with accounts of the yeti will know that this theory just will not hold water. The bear confusion may arise from the term dzu-teh which is used to describe large or hulking creatures and is applied to both the yeti and the brown bear. In a similar vein he says that most early Sasquatch reports were of creatures smaller than men, which is quite untrue, and gives credence to the 'Jacko' story of 1884 where a small Bigfoot like creature was supposedly captured by railway workers. The story was almost certainly a newspaper hoax.

These minor quibbles aside, Forth's work is a really quite staggering piece of research, and an invaluable tool for reference on this little studied subject. **RICHARD FREEMAN**

Citro, Joseph A, Bisset, Stephen R
The Vermont Monster Guide
University Press of New England

If you're a comic book geek like me (I recently sold my massive Batman collection to finance a trip to Sumatra) you will know Stephen R Bisset as the superb artist on DCs infamous *Swamp Thing*. And what better artist to illustrate Joseph A Citro's wonderful guide to the monsters of the Green Mountain State! It's not the Congo or the Amazon but Vermont seems to have more than its share of monsters both historical and modern.

The book is not detailed but it's not meant to be. It very much picture led with Bisset's moody and evocative illustrations. Having said that, the book is no coffee table edition. Much information in here was totally new to me. In focusing on one state Citro can tell stories that a larger book, covering a wider area might miss.

And what strange stories they are. Alongside well known monsters such as Champ, thunderbirds, ABCs and Bigfoot you have oddities like the Pigman of Northfield, a sort of humanoid beast with a pig's head, giant rabbits, a 300 lb leech and a sighting of a creature resembling the Mexican dragon god Quetzalcoatl.

It might be a nice project to create a book like this for each and every state. **RICHARD FREEMAN**

87

Deep in a cave beneath Loch Ness lives a strange figure who steals ideas from other magazines and then somehow makes them his own.

Jon and Corinna
GONE TO TEXAS

Howdy Y'all

No doubt, most of next issue's amusing bits and bobs will be drawn from what will happen at the 11th Weird Weekend, but for this issue I shall be casting my spotlight upon the events that transpired when Mr & Mrs D went to the Lone Star State.

Those of you with long memories will remember how, in 1998, Jon complained that Graham had 'gone native'. It is true, Jon insists, he did. However, he draws a convenient veil over what happened when he and Corinna went to Texas earlier in the year. As the evidence in this photograph proves, they 'went native' to a far more ludicrous extent. Those who live in glass houses, Jon......

Still on the subject of Texas, readers of Jon's first volume of trans-Atlantic exploits - *Only Fools and Goatsuckers* (2002) - which was written about the aforementioned 1998 trips to Mexico and Puerto Rico may remember another incident involving Graham and an unfortunately named brand of soft drink.

Much to Jon's joy he found the self-same brand on sale in San Antonio and to his greater joy upon their arrival back in England he found a frame from the original movie.

THE SYCOPHANT

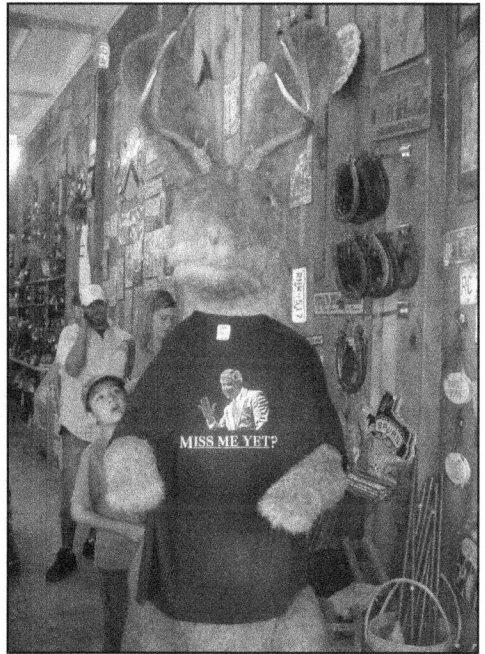

Although we at the CFZ HQ do not echo the sentiments, it was certainly amusing that when Jon and Corinna visited Jackalope Joe's, the shop in downtown San Antonio which sells - you've guessed it - jackalopes, to find one (soft toy) over 8 foot in height, wearing a T-shirt with a picture of George W Bush and the caption "missed me yet?"

This would hardly be likely to happen in the UK because, as the recent election shows, nobody is impressed enough with any of the main parties to give them a mandate, and most contemporary British politicians, being chosen by focus groups who are determined not to upset anyone, are so insipid, as to be unrecognisable on or off a t-shirt.

I think that this photograph taken by Corinna of the bumper sticker on SMiles Lewis of the Anomaly Archives in Austin's car sums up the current state of British politics pretty damn well.

If anyone wants to know more about where the Sycophant's "politics are" I will only say "hey now baby, get into my big black car, and check out the last verse of 'White Punks on Hope'."

THE CENTRE FOR FORTEAN ZOOLOGY

So, what is the Centre for Fortean Zoology?

We are a non profit-making organisation founded in 1992 with the aim of being a clearing house for information, and coordinating research into mystery animals around the world. We also study out of place animals, rare and aberrant animal behaviour, and Zooform Phenomena; little-understood "things" that appear to be animals, but which are in fact nothing of the sort, and not even alive (at least in the way we understand the term).

Why should I join the Centre for Fortean Zoology?

Not only are we the biggest organisation of our type in the world, but - or so we like to think - we are the best. We are certainly the only truly global Cryptozoological research organisation, and we carry out our investigations using a strictly scientific set of guidelines. We are expanding all the time and looking to recruit new members to help us in our research into mysterious animals and strange creatures across the globe. Why should you join us? Because, if you are genuinely interested in trying to solve the last great mysteries of Mother Nature, there is nobody better than us with whom to do it.

What do I get if I join the Centre for Fortean Zoology?

For £12 a year, you get a four-issue subscription to our journal *Animals & Men*. Each issue contains 60 pages packed with news, articles, letters, research papers, field reports, and even a gossip column! The magazine is A5 in format with a full colour cover. You also have access to one of the world's largest collections of resource material dealing with cryptozoology and allied disciplines, and people from the CFZ membership regularly take part in fieldwork and expeditions around the world.

How is the Centre for Fortean Zoology organised?

The CFZ is managed by a three-man board of trustees, with a non-profit making trust registered with HM Government Stamp Office. The board of trustees is supported by a Permanent Directorate of full and part-time staff, and advised by a Consultancy Board of specialists - many of whom are world-renowned experts in their particular field. We have regional representatives across the UK, the USA, and many other parts of the world, and are affiliated with other organisations whose aims and protocols mirror our own.

I am new to the subject, and although I am interested I have little practical knowledge. I don't want to feel out of my depth. What should I do?

Don't worry. We were *all* beginners once. You'll find that the people at the CFZ are friendly and approachable. We have a thriving forum on the website which is the hub of an ever-growing electronic community. You will soon find your feet. Many members of the CFZ Permanent Directorate started off as ordinary members, and now work full-time chasing monsters around the world.

I have an idea for a project which isn't on your website. What do I do?

Write to us, e-mail us, or telephone us. The list of future projects on the website is not exhaustive. If you have a good idea for an investigation, please tell us. We may well be able to help.

How do I go on an expedition?

We are always looking for volunteers to join us. If you see a project that interests you, do not hesitate to get in touch with us. Under certain circumstances we can help provide funding for your trip. If you look on the future projects section of the website, you can see some of the projects that we have pencilled in for the next few years.

In 2003 and 2004 we sent three-man expeditions to Sumatra looking for Orang-Pendek - a semi-legendary bipedal ape. The same three went to Mongolia in 2005. All three members started off merely subscribers to the CFZ magazine.

Next time it could be you!

Project Kerinci, Sumatra - 2003
In search of the bipedal ape Orang Pendek

How is the Centre for Fortean Zoology funded?

We have no magic sources of income. All our funds come from donations, membership fees, works that we do for TV, radio or magazines, and sales of our publications and merchandise. We are always looking for corporate sponsorship, and other sources of revenue. If you have any ideas for fund-raising please let us know. However, unlike other cryptozoological organisations in the past, we do not live in an intellectual ivory tower. We are not afraid to get our hands dirty, and furthermore we are not one of those organisations where the membership have to raise money so that a privileged few can go on expensive foreign trips. Our research teams, both in the UK and abroad, consist of a mixture of experienced and inexperienced personnel. We are truly a community, and work on the premise that the benefits of CFZ membership are open to all.

What do you do with the data you gather from your investigations and expeditions?

Reports of our investigations are published on our website as soon as they are available. Preliminary reports are posted within days of the project finishing.

Each year we publish a 200 page yearbook containing research papers and expedition reports too long to be printed in the journal. We freely circulate our information to anybody who asks for it.

Is the CFZ community purely an electronic one?

No. Each year since 2000 we have held our annual convention - the *Weird Weekend* - in Exeter. It is three days of lectures, workshops, and excursions. But most importantly it is a chance for members of the CFZ to meet each other, and to talk with the members of the permanent directorate in a relaxed and informal setting and preferably with a pint of beer in one hand. Since 2006 - the *Weird Weekend* has been bigger and better and held on the third weekend in August in the idyllic rural location of Woolsery in North Devon.

Since relocating to North Devon in 2005 we have become ever more closely involved with other community organisations, and we hope that this trend will continue. We also work closely with Police Forces across the UK as consultants for animal mutilation cases, and we intend to forge closer links with the coastguard and other community services. We want to work closely with those who regularly travel into the Bristol Channel, so that if the recent trend of exotic animal visitors to our coastal waters continues, we can be out there as soon as possible.

We are building a Visitor's Centre in rural North Devon. This will not be open to the general public, but will provide a museum, a library and an educational resource for our members (currently over 400) across the globe. We are also planning a youth organisation which will involve children and young people in our activities.

Apart from having been the only Fortean Zoological organisation in the world to have consistently published material on all aspects of the subject for over a decade, we have achieved the following concrete results:

- Disproved the myth relating to the headless so-called sea-serpent carcass of Durgan beach in Cornwall 1975
- Disproved the story of the 1988 puma skull of Lustleigh Cleave
- Carried out the only in-depth research ever into the mythos of the Cornish Owlman
- Made the first records of a tropical species of lamprey
- Made the first records of a luminous cave gnat larva in Thailand
- Discovered a possible new species of British mammal - the beech marten
- In 1994-6 carried out the first archival fortean zoological survey of Hong Kong
- In the year 2000, CFZ theories were confirmed when an new species of lizard was added to the British list
- Identified the monster of Martin Mere in Lancashire as a giant wels catfish
- Expanded the known range of Armitage's skink in the Gambia by 80%
- Obtained photographic evidence of the remains of Europe's largest known pike
- Carried out the first ever in-depth study of the *ninki-nanka*
- Carried out the first attempt to breed Puerto Rican cave snails in captivity
- Were the first European explorers to visit the `lost valley` in Sumatra
- Published the first ever evidence for a new tribe of pygmies in Guyana
- Published the first evidence for a new species of caiman in Guyana
- Filmed unknown creatures on a monster-haunted lake in Ireland for the first time
- Had a sighting of orang pendek in Sumatra in 2009
- Published some of the best evidence ever for the almasty in southern Russia

- 1998 Puerto Rico, Florida, Mexico *(Chupacabras)*
- 1999 Nevada *(Bigfoot)*
- 2000 Thailand *(Giant snakes called nagas)*
- 2002 Martin Mere *(Giant catfish)*
- 2002 Cleveland *(Wallaby mutilation)*
- 2003 Bolam Lake *(BHM Reports)*
- 2003 Sumatra *(Orang Pendek)*
- 2003 Texas *(Bigfoot; giant snapping turtles)*
- 2004 Sumatra *(Orang Pendek; cigau, a sabre-toothed cat)*
- 2004 Illinois *(Black panthers; cicada swarm)*
- 2004 Texas *(Mystery blue dog)*
- Loch Morar *(Monster)*
- 2004 Puerto Rico *(Chupacabras; carnivorous cave snails)*
- 2005 Belize *(Affiliate expedition for hairy dwarfs)*
- 2005 Loch Ness *(Monster)*
- 2005 Mongolia *(Allghoi Khorkhoi aka Mongolian death worm)*
- 2006 Gambia *(Gambian sea monster , Ninki Nanka and Armitage's skink*
- 2006 Llangorse Lake *(Giant pike, giant eels)*
- 2006 Windermere *(Giant eels)*
- 2007 Coniston Water *(Giant eels)*
- 2007 Guyana *(Giant anaconda, didi, water tiger)*
- 2008 Russia *(Almasty)*
- 2009 Sumatra *(Orang pendek)*
- 2009 Republic of Ireland *(Lake Monster)*
- 2010 Texas *(Blue dogs)*

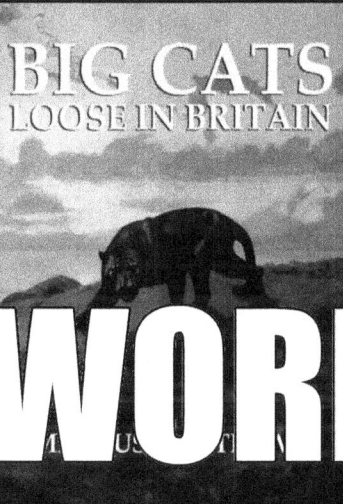

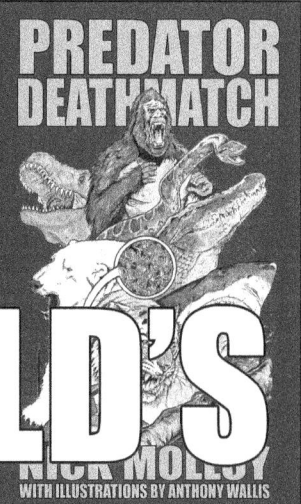

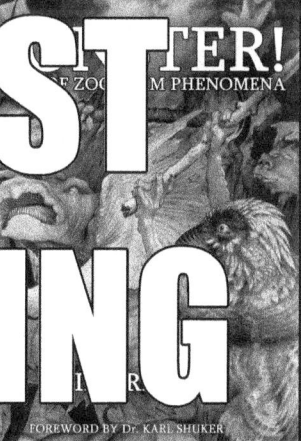

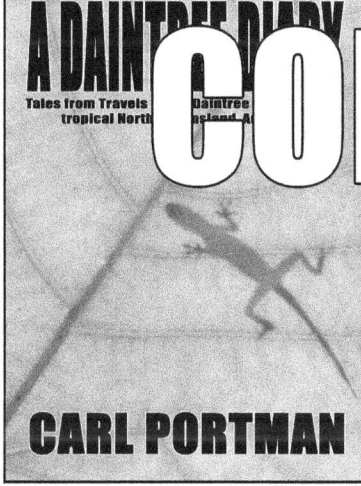

THE WORLD'S WEIRDEST PUBLISHING COMPANY

HOW TO START A PUBLISHING EMPIRE

Unlike most mainstream publishers, we have a non-commercial remit, and our mission statement claims that "we publish books because they deserve to be published, not because we think that we can make money out of them". Our motto is the Latin Tag "Pro bona causa facimus" (we do it for good reason), a slogan taken from a children's book `The Case of the Silver Egg` by the late Desmond Skirrow.

WIKIPEDIA: "The first book published was in 1988. `Take this Brother may it Serve you Well` was a guide to Beatles bootlegs by Jonathan Downes. It sold quite well, but was hampered by very poor production values, being photocopied, and held together by a plastic clip binder. In 1988 A5 clip binders were hard to get hold of, so the publishers took A4 binders and cut them in half with a hacksaw. It now reaches surprisingly high prices second hand.

The production quality improved slightly over the years, and after 1999 all the books produced were ringbound with laminated colour covers. In 2004, however, they signed an agreement with LightningSource, and all books are now produced perfect bound, with full colour covers."

Until 2010 all our books, the majority of which are/were on the subject of mystery animals and allied disciplines, were published by `CFZ Press`, the publishing arm of the Centre for Fortean Zoology (CFZ), and we urged our readers and followers to draw a discreet veil over the books that we published that were completely off topic to the CFZ.

However, in 2010 we decided that enough was enough and launched a second imprint, `Fortean Words` which aims to cover a wide range of non animal-related esoteric subjects. Other imprints will be launched as and when we feel like it, however the basic ethos of the company remains the same: Our job is to publish books and magazines that we feel are worth publishing, whether or not they are going to sell. Money is, after all - as my dear old Mama once told me - a rather vulgar subject, and she would be rolling in her grave if she thought that her eldest son was somehow in `trade`.

Luckily, so far our tastes have turned out not to be that rarified after all, and we have sold far more books than anyone ever thought that we would, so there is a moral in there somewhere…

Jon Downes,
Woolsery, North Devon
July 2010

CFZ PRESS

Other Books in Print

The Mystery Animals of Ireland by Gary Cunningham and Ronan Coghlan
Monsters of Texas by Gerhard, Ken
The Great Yokai Encyclopaedia by Freeman, Richard
NEW HORIZONS: Animals & Men *issues 16-20 Collected Editions Vol. 4* by Downes, Jonathan
A Daintree Diary -
Tales from Travels to the Daintree Rainforest in tropical north Queensland, Australia by Portman, Carl
Strangely Strange but Oddly Normal by Roberts, Andy
Centre for Fortean Zoology Yearbook 2010 by Downes, Jonathan
Predator Deathmatch by Molloy, Nick
Star Steeds and other Dreams by Shuker, Karl
CHINA: A Yellow Peril? by Muirhead, Richard
Mystery Animals of the British Isles: The Western Isles by Vaudrey, Glen
Giant Snakes - Unravelling the coils of mystery by Newton, Michael
Mystery Animals of the British Isles: Kent by Arnold, Neil
Centre for Fortean Zoology Yearbook 2009 by Downes, Jonathan
CFZ EXPEDITION REPORT: Russia 2008 by Richard Freeman *et al*, Shuker, Karl (fwd)
Dinosaurs and other Prehistoric Animals on Stamps - A Worldwide catalogue by Shuker, Karl P. N
Dr Shuker's Casebook by Shuker, Karl P.N
The Island of Paradise - chupacabra UFO crash retrievals,
and accelerated evolution on the island of Puerto Rico by Downes, Jonathan
The Mystery Animals of the British Isles: Northumberland and Tyneside by Hallowell, Michael J
Centre for Fortean Zoology Yearbook 1997 by Downes, Jonathan (Ed)
Centre for Fortean Zoology Yearbook 2002 by Downes, Jonathan (Ed)
Centre for Fortean Zoology Yearbook 2000/1 by Downes, Jonathan (Ed)
Centre for Fortean Zoology Yearbook 1998 by Downes, Jonathan (Ed)
Centre for Fortean Zoology Yearbook 2003 by Downes, Jonathan (Ed)
In the wake of Bernard Heuvelmans by Woodley, Michael A
CFZ EXPEDITION REPORT: Guyana 2007 by Richard Freeman *et al*, Shuker, Karl (fwd)

Centre for Fortean Zoology Yearbook 1999 by Downes, Jonathan (Ed)
Big Cats in Britain Yearbook 2008 by Fraser, Mark (Ed)
Centre for Fortean Zoology Yearbook 1996 by Downes, Jonathan (Ed)
THE CALL OF THE WILD - Animals & Men issues 11-15
Collected Editions Vol. 3 by Downes, Jonathan (ed)
Ethna's Journal by Downes, C N
Centre for Fortean Zoology Yearbook 2008 by Downes, J (Ed)
DARK DORSET -Calendar Custome by Newland, Robert J
Extraordinary Animals Revisited by Shuker, Karl
MAN-MONKEY - In Search of the British Bigfoot by Redfern, Nick
Dark Dorset Tales of Mystery, Wonder and Terror by Newland, Robert J and Mark North
Big Cats Loose in Britain by Matthews, Marcus
MONSTER! - The A-Z of Zooform Phenomena by Arnold, Neil
The Centre for Fortean Zoology 2004 Yearbook by Downes, Jonathan (Ed)
The Centre for Fortean Zoology 2007 Yearbook by Downes, Jonathan (Ed)
CAT FLAPS! Northern Mystery Cats by Roberts, Andy
Big Cats in Britain Yearbook 2007 by Fraser, Mark (Ed)
BIG BIRD! - Modern sightings of Flying Monsters by Gerhard, Ken
THE NUMBER OF THE BEAST - Animals & Men issues 6-10
Collected Editions Vol. 1 by Downes, Jonathan (Ed)
IN THE BEGINNING - Animals & Men issues 1-5 Collected Editions Vol. 1 by Downes, Jonathan
STRENGTH THROUGH KOI - They saved Hitler's Koi and other stories by Downes, Jonathan
The Smaller Mystery Carnivores of the Westcountry by Downes, Jonathan
CFZ EXPEDITION REPORT: Gambia 2006 by Richard Freeman *et al*, Shuker, Karl (fwd)
The Owlman and Others by Jonathan Downes
The Blackdown Mystery by Downes, Jonathan
Big Cats in Britain Yearbook 2006 by Fraser, Mark (Ed)
Fragrant Harbours - Distant Rivers by Downes, John T
Only Fools and Goatsuckers by Downes, Jonathan
Monster of the Mere by Jonathan Downes
Dragons:More than a Myth by Freeman, Richard Alan
Granfer's Bible Stories by Downes, John Tweddell
Monster Hunter by Downes, Jonathan

Fortean Words

The Centre for Fortean Zoology has for several years led the field in Fortean publishing. CFZ Press is the only publishing company specialising in books on monsters and mystery animals. CFZ Press has published more books on this subject than any other company in history and has attracted such well known authors as Andy Roberts, Nick Redfern, Michael Newton, Dr Karl Shuker, Neil Arnold, Dr Darren Naish Jon Downes, Ken Gerhard and Richard Freeman.

Now CFZ Press are launching a new imprint. Fortean Words is a new line of books dealing with Fortean subjects other than cryptozoology, which is - after all - the subject the CFZ are best known for. Fortean Words is being launched with a spectacular multi-volume series called *Haunted Skies* which covers British UFO sightings between 1940 and 2010. Former policeman John Hanson and his long-suffering partner Dawn Holloway have compiled a peerless library of sighting reports, many that have been made public before.

Other forthcoming books include a look at the Berwyn Mountains UFO case by renowned Fortean Andy Roberts and a series of books by transatlantic research Nick Redfern.

CFZ Press are dedicated to maintaining the fine quality of their works with Fortean Words. New authors tackling new subjects will always be encouraged, and we hope that our books will continue to be as ground breaking and popular as ever.